S0-DUU-622

A SIMPLE HISTORY OF
THE STEAM ENGINE

A SIMPLE HISTORY OF

THE STEAM ENGINE

by J. D. Storer

Original drawings by H. Fernandez

JOHN BAKER PUBLISHER LONDON 1969

SBN 212.98356.3

First Published by John Baker (Publishers) Ltd
5 Royal Opera Arcade, London S.W.1
Printed in England at The Curwen Press, London E.13

Contents

List of Diagrams

FIG

Fig

List of Illustrations

Acknowledgements

I wish to thank all my friends and colleagues for the help they have given me during the writing of this book. In particular my thanks are due to my wife and Mrs. B. Sugden for their work in the preparation of the manuscript.

Illustrations are a very important part of this type of book and I am very grateful to Mr. Fernandez for his meticulous work on the line drawings. In certain cases I have used photographs and drawings from early books and I am indebted to the Director of the Royal Scottish Museum for permission to use Museum negatives.

The author and publishers would like to thank all the organisations who supplied the photographs from which we made our selection. It was impossible to use all the available material and an endeavour was made to use less well-known illustrations. We are very grateful to the copyright holders for permission to reproduce the photographs listed below:

Plates 1–4, 6–11, 13, 14, 20, 21, 25, 30, 31, 35, 37. Royal Scottish Museum, Edinburgh (Crown Copyright).
Plates 12, 22. The Art Gallery and Museum, Glasgow.
Plates 23, 40. English Electric Co. Ltd., Rugby.
Plate 26. Babcock and Wilcox Ltd., London.
Plate 27. Forth Ports Authority.
Plate 28. Trevor Rees Esq.
Plates 29, 34. J. D. Storer Esq.
Plate 32. Wm. Dobson (Edin.) Ltd.
Plate 36. Museum of Science & Engineering, Newcastle.
Plate 41. C. A. Parsons & Co. Ltd., Newcastle.
Plate 42. Upper Clyde Shipbuilders Ltd.
Plate 44. United Kingdom Atomic Energy Authority.

J.D.S.

Bibliography

E. L. Ahrons, *The British Steam Railway Locomotive*, (*Vol.* 1, 1825–1925), Ian Allan, London, 1927 (reprint 1966).
M. Archer, *William Hedley*, Newcastle-upon-Tyne, 1882.
W. A. Baker, *From Paddle Steamer to Nuclear Ship*, C. A. Watts, London, 1965.
R. H. Clark, *The Development of the English Traction Engine*, Goose and Son.
H. W. Dickinson, *A Short History of the Steam Engine*, The University Press, Cambridge, 1938.
H. W. Dickinson and Rhys Jenkins, *James Watt and the Steam Engine*, Oxford University Press, 1927.
J. Farey, *A Treatise on the Steam Engine*, Longmans, London, 1827.
R. J. Law, *The Steam Engine*, Science Museum/HMSO, London, 1965.
O. S. Nock, *British Steam Railway Locomotives 1925–1965*, Ian Allan, London, 1966.
L. T. C. Rolt, *James Watt*, B. T. Batsford, London, 1962.
L. T. C. Rolt, *Thomas Newcomen*, David & Charles, Dawlish, 1963.
C. Singer et al., *A History of Technology Vols. 1 to 5*, Oxford University Press, c.1955.
R. Stuart, *Descriptive History of the Steam Engine*, John Chidley, London, 1831.
Wm. Symington, *Steam Navigation*, Falkirk 1829 (reprint 1863).
R. H. Thurston, *A History of the Growth of the Steam Engine*, C. Kegan Paul, London, 1879.
G. Watkins, *The Stationary Steam Engine*, David & Charles, Newton Abbot, 1968.
Various Authors, *The Edinburgh Encyclopaedia*, W. Blackwood, Edinburgh, c.1830.
Various Authors, *Engineering Heritage Vols. 1 & 2*, Institution of Mechanical Engineers/Heineman, London, 1963 & 1966.
Various Authors, *Historical Surveys & Descriptive Catalogues*, Science Museum/HMSO, London.
Various Authors (Periodicals), *Industrial Archaeology*, *Transport History*, *The Engineer*, *Engineering*, *Model Engineer*.

CHAPTER I

An Introduction to the Steam Engine

And other sources of power

The Moon is now only a few days away by rocket-powered spacecraft and this exciting new mode of transport stirs the imagination of all. To the children of today the ambition to be an astronaut is just as natural as was their fathers' desire to be an aeroplane pilot, or their grandfathers' to become an engine-driver. The power from a rocket motor makes the effort of a steam locomotive appear insignificant, yet in their day these very locomotives caused people to gaze in awe. Despite their differences, the rocket, aeroplane and locomotive have one thing in common: their power is obtained by burning fuel, or in other words they are all heat engines.

The now almost discarded steam engine was clearly a heat engine because a fire heating its boiler could easily be seen. In contrast, the petrol engine's use of heat is less obvious because it is an 'internal-combustion' design which burns fuel inside the cylinder instead of in an external boiler. This book aims to tell the story of the steam engine, but this story cannot be told fairly unless some of the rival sources of power are also mentioned. After all, if the petrol engine had not been invented our present-day motor cars might have had steam engines.

One of man's greatest achievements was the conversion of heat energy into useful power, since previously he had had to rely on natural sources of power. Muscle, wind and water power served him well for many centuries, but they had their limitations; muscles become tired, the wind drops and fast-flowing rivers are not always conveniently sited.

The only source of power known in prehistoric days was man's own muscles. If primitive people wished to travel they walked or paddled a dug-out canoe. In order to grow crops they tilled the soil, sowed, harvested and ground the corn, all by hand. Eventually men tamed the more intelligent animals, such as the horse and the ox, thus bringing into use a new source of muscle power. The superior strength of these animals made them very useful for carrying goods, pulling a plough and even driving simple machinery. For instance, pumps of various types were sometimes used to supply water for the irrigation of cultivated land, particularly in the warm countries around the Mediterranean Sea. This region was the centre of several early civilizations, and it is interesting to note that the superior strength of the animal did not displace the more controllable muscles of men—who were often slaves. The Seven Wonders of the World, including the Egyptian Pyramids, were built almost entirely by the muscles of men.

At sea, the Romans pressed into service gangs of slaves to power the oars of their galleys, but they also utilized the power of the wind. Sailing ships had been used for many centuries; the Phoenicians in about 1000 B.C. developed wooden sailing ships which later reached as far as Britain. Despite the fickle nature of the wind, in both strength and direction, it supplied the power for the majority of ships until the end of the last century.

On land, the power of the wind was harnessed also by a sail, but in this case the sail went round and round on the arms of a windmill. Strangely enough the windmill is a relatively new invention compared with the sailing ship or waterwheel, for the first reliable descriptions date from about A.D. 1000. Windmills were used primarily to grind corn and pump water, this latter task being still performed by windmills on some farms.

Probably man's first attempt to use a source of power, other than muscles, was when he sat on a log and floated downstream in the current of a river. As a means of transport, drifting along had its limitations, but the power of a fast-moving river must have been obvious. It is not known who invented the waterwheel but one was mentioned in a poem written almost 100 years B.C. Many of the early water mills were used for grinding corn into flour and they were built with their spindles in a vertical position driving a circular mill-stone directly above the waterwheel. As the wheel was lying flat the stream had to be channelled in such a way that it flowed past one side of the wheel only, so striking the blades and turning the wheel.

A Roman engineer called Pollio Vitruvius, who lived in the first century B.C., described a waterwheel mounted on a horizontal spindle – a layout well known even in modern times. Since the mill-stones were still required in the same position, gear-wheels had to be used to connect the waterwheel to the rotating stones. The use of gears, however, enabled the mill-stones to be driven faster than the slow-moving waterwheel. Very primitive but none the less effective gear-wheels had been in use for some 200 years before Vitruvius and, of course, they became a vital part of later engines.

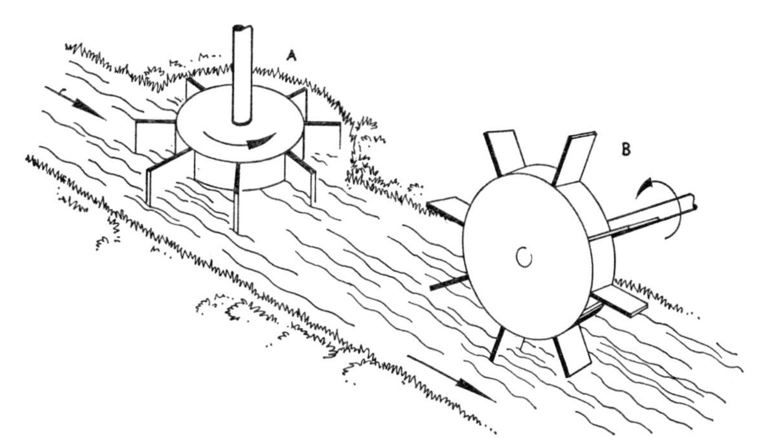

FIG 1 *Simple waterwheels.* A *Early type with vertical spindle.* B *More usual layout, described by Vitruvius.*

Water is still a very important source of power even today, for in mountainous countries electricity is generated in hydro-electric power-stations. Water from a lake or reservoir, situated high in the mountains, is piped down the mountainside and by the time it reaches the power-station it is moving very rapidly. Instead of turning a simple waterwheel it drives more complicated and efficient water turbines, which in turn drive electrical generators.

This use of water power also serves as a good example to illustrate the two types of energy. The water high in the mountains is said to have potential or stored-up energy, while the fast-moving water which drives the turbines has kinetic or moving energy. In order to obtain power from any stored-up source of energy it has to be converted into kinetic energy of some kind. Examples of this idea can be found in many places. A clock spring when wound up has potential energy: the moving parts have kinetic energy. A bow drawn ready to shoot an arrow has potential energy: once released, the arrow has kinetic energy.

A piece of coal does not look like a store of energy, yet in reality it has considerable potential energy, for heat is a form of energy and by burning coal, of course, heat is obtained. Releasing this energy is one thing, but harnessing it presents quite another problem. Primitive men learnt how to make a fire using wood as fuel, but little did they realize the power being wasted. After all, the only indication of the kinetic energy contained in the hot air rising above the fire was smoke and a few pieces of ash floating upwards.

Heat engines have one thing in common – they all derive their power from potential energy stored in fuels. In most cases the fuel is burnt to produce heat energy which, in turn, is converted into mechanical power. This final transformation proved to be a complicated process, which required the solution of many scientific and engineering problems before a practical heat engine emerged. However, the problems were overcome and in 1712 Thomas Newcomen built the first practical steam engine. Heat from a fire below a boiler converted water into steam which was then used to drive a sliding piston inside a cylinder. If this type of engine, known as a reciprocating engine, is compared with a bicycle pump, then the cylinder would correspond to the main body of the pump, the piston to the washer inside the pump, and the piston rod to the pump's handle.

The piston of a reciprocating engine – as its name suggests – moves to and fro; thus, if the engine is to turn a wheel or shaft, some mechanism is required to convert reciprocating motion into rotary motion. The simplest way of doing this is to introduce a crank and connecting rod. Perhaps the most common example of a crank is the bicycle pedal, but in this instance the connecting rods, represented by the rider's legs, bend instead of reciprocating. In a reciprocating engine the piston rod – or in some cases the piston itself – is linked

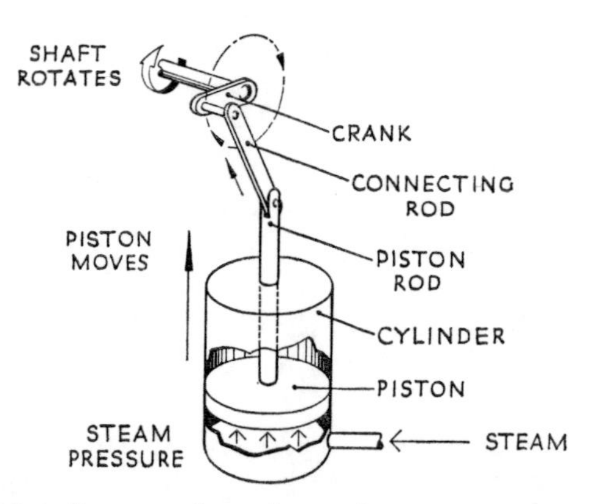

FIG 2 *A simplified diagram of a reciprocating steam engine.*

by a connecting rod to a crankshaft which rotates and drives wheels or machinery.

Anyone who has ever seen a large, gleaming steam engine pounding round and round cannot fail to sense the power of such an engine, yet it all depends on a relatively insignificant pipe bringing high-pressure steam from a boiler. Often the size of the boiler, or boilers, dwarfs the engine as in the case of a steam railway locomotive. The main bulk of a locomotive above its wheels is taken up by the boiler, while the cylinders of its reciprocating engine can often be seen over the wheels of the front bogie – minute by comparison. Stationary steam engines, which powered many factories, were supplied with steam from a central boiler-house. But in both the locomotive and the factory, the burning of fuel to generate heat was separated from the working parts of the engine. However, this is not the only way to make a reciprocating heat engine.

Fuel can be burnt inside the cylinder of an engine, in which case the high-pressure gases so formed act directly on the piston. As combustion takes place inside the engine this is naturally called an

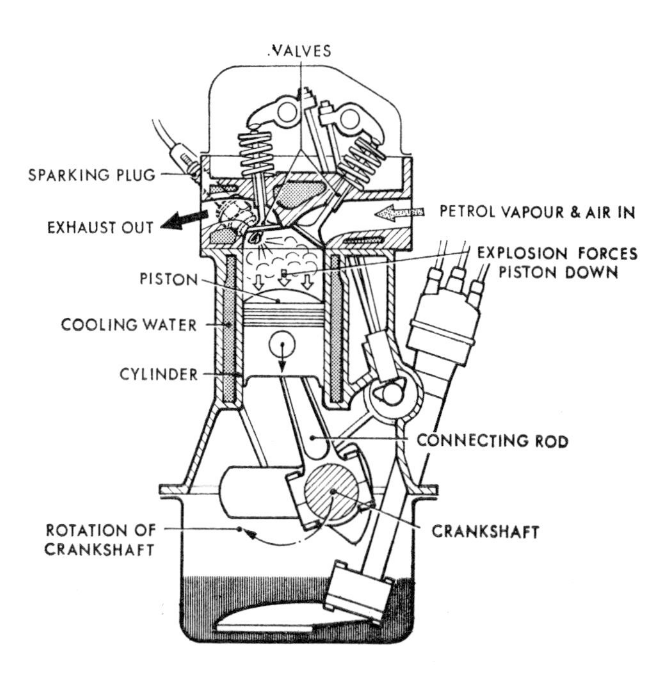

FIG 3 *An internal-combustion engine, used to power a motor car.*

internal-combustion engine. The fuel can be petrol, paraffin, diesel oil, coal gas or one of several other inflammable substances. The petrol engine, as used in cars, is the most widely known internal-combustion engine, followed by the diesel engine, which is more suitable for larger installations including buses, trains and ships. Tractors often burn cheap paraffin but gas engines are rarely seen in modern times, although they played an important rôle in the early days of the internal-combustion engine.

Pistons moving backwards and forwards with long connecting rods and heavy crankshafts are not the ideal mechanism for producing rotation. What could be more simple than waterwheels or windmills? Place them in a stream of water or wind respectively and they rotate without any complicated mechanism. But they have a major snag – the unreliability of their source of power. Now if the water or wind could be replaced by a man-made stream of steam, or hot gases, an ideal rotary engine would result. This type of heat engine is known as a turbine and, like the reciprocating engine, it has two branches, steam and internal-combustion, although the latter is usually called a gas turbine.

The simplest form of steam turbine is closely related to the waterwheel. High-pressure steam from a boiler is passed through a nozzle, from which it emerges as a high-speed jet, its pressure (or potential) energy having been converted into kinetic energy. The jet of steam is directed on to carefully shaped blades mounted on a wheel and the impact of the jet drives the wheel round at a high speed. This is known as an 'impulse' turbine.

When a gun is fired, the bullet moves in one direction while the gun recoils in the opposite direction. The same effect is achieved when a jet of steam leaves a nozzle at high speed. If the recoil of the nozzle is harnessed to drive a wheel, then the result is a 'reaction' turbine.

Most steam turbines are, in fact, a combination of both the 'impulse' and 'reaction' types. High-pressure steam from a boiler passes through a series of 'windmills' or, more precisely, rotating discs with 'blades' instead of arms with sails. Between each rotating disc, with its numerous blades, there is a similar fixed disc which ensures that the steam is directed in the right direction for the next moving disc. As the steam expands on its journey through the stages of the turbine so the size of the blades increases accordingly. When the steam finally emerges, most of its pressure energy has been passed to the rotating discs as kinetic energy. To be efficient steam turbines have to run at very high speeds which makes them

PLATE 1 *The Rolls-Royce* Derwent *of 1945—cut away to show important features: from right to left: engine accessories, air intake (covered by wire mesh), air compressor, combustion chambers ('cans'), turbine and jet pipe.*

PLATE 2 *A model of Nicholas Cugnot's road vehicle or* fardier *of 1769–70.*

PLATE 3 *A working model of a rotative Newcomen-type engine. The actual engine was built at Farme Colliery in Scotland where it was used to wind the cage up and down the mine shaft, also for pumping water during the night. It was in service from 1810 to 1915 and is now preserved in the Glasgow Museum.*

PLATE 4 *A model of James Watt's* Lap *engine, 1788. A typical example of a Watt rotative beam engine with: separate condenser, double-acting piston, parallel motion, sun-and-planet gearing and governor.*

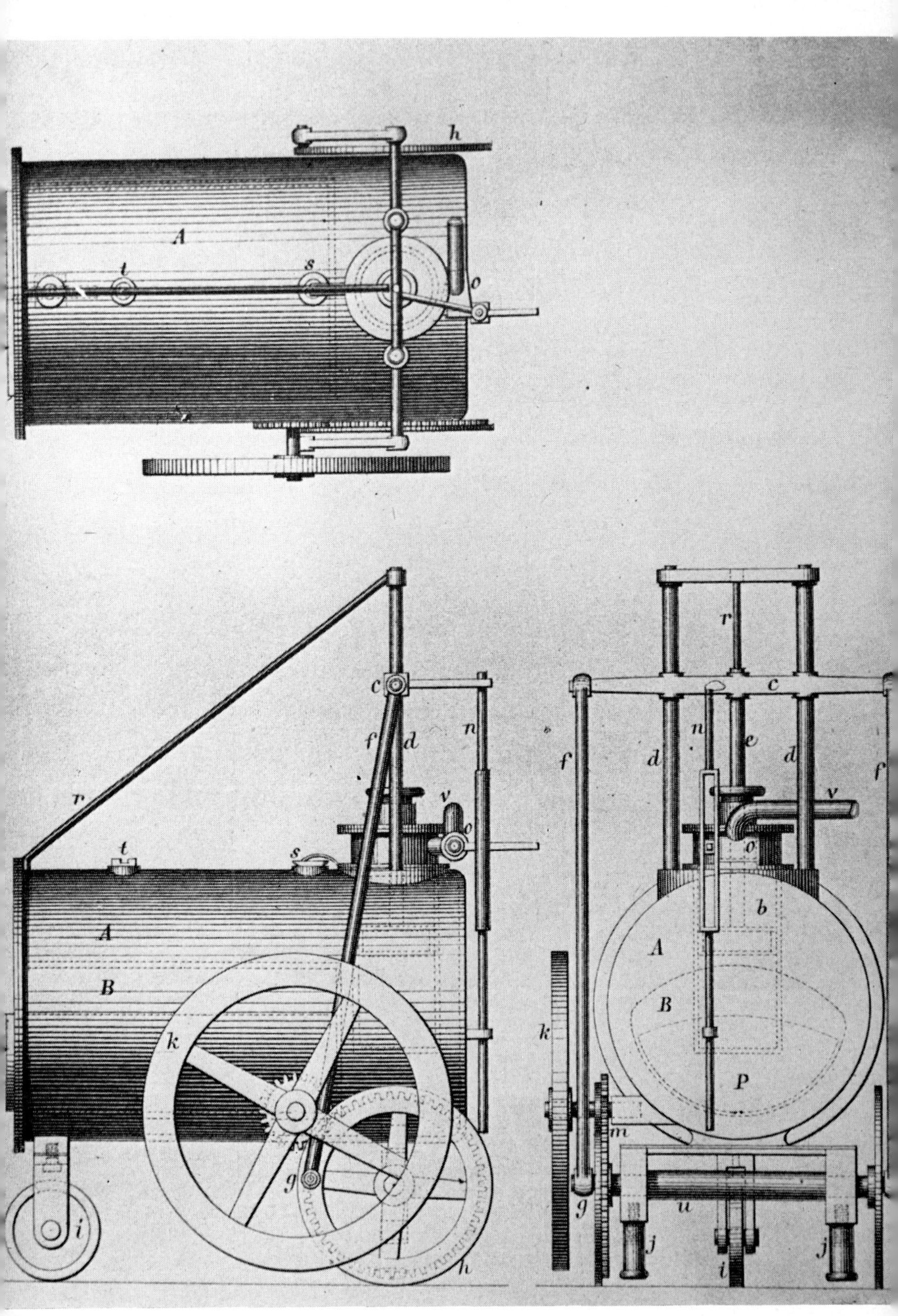

PLATE 5 *One of Trevithick's models of the late 1790's. The vertical cylinder is mounted inside the boiler. The piston rod* e *is attached to the crosshead* c *which is guided along a straight line by sliding along rods* d. *Connecting rods* f *give a direct drive to the crank* g.

ideal for driving electrical generators, but in order to adapt them for a ship, a gear-box is needed to reduce the speed of rotation to suit a propeller.

Turbines are a very efficient method of converting steam pressure into mechanical power, but the overall efficiency includes the generating of heat to produce steam – a boiler, for example. The early boilers burned coal and were very simple in design, but not very efficient. More complicated designs, absorbing a greater amount of heat from a fire, increased the efficiency considerably while mechanical stoking reduced the manual work. Other fuels have also been tried including gas and oil, the latter proving to be very successful for ships. Steam turbines fed by oil-fired boilers are still a successful

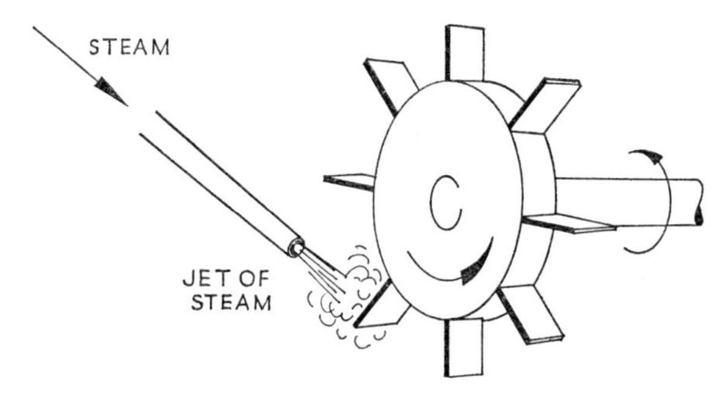

FIG 4 *A simple steam turbine of the 'impulse' type*

form of power for large fast ships, such as the *Queen Elizabeth II.*

Although steam reciprocating engines are rapidly disappearing, steam turbines are resisting the threat from the internal-combustion engine – in particular the diesel. In fact, they are enjoying a new lease of life due to one of the greatest scientific discoveries of all time, atomic or nuclear energy. For many years scientists were aware that a great quantity of potential energy was stored inside atoms; the problem was how to release this energy. It was ultimately released in the atomic bomb with devastating effect. The next stage was to control this source of power and utilize it for more peaceful purposes.

The key to the problem was the building of a 'reactor' in which the central core or 'nucleus' of certain atoms could be split, under control, thus releasing large quantities of heat. This nuclear heat is now used in boilers to raise steam for the turbines of a number of electrical power stations, ships and submarines. In the future, nuclear 'fuel' may well supersede coal and oil, but at the present time the size of the installation limits its use.

A very compact form of turbine can be obtained by adopting the internal-combustion principle and burning fuel, usually a type of paraffin, inside the engine. Hot gases at high pressure result from the combustion and these turn the discs of a turbine – hence the name 'gas turbine'. To achieve greater power and efficiency the fuel is burnt in compressed air, and so gas turbines are fitted with an air compressor. This is often a series of rotating discs with blades gradually reducing in size – rather like a turbine in reverse. Power to drive the compressor comes from the turbine by means of a shaft along the centre of the engine. Sometimes this same shaft is used to drive a propeller or wheels, while industrial gas turbines drive electrical generators or machinery. But the most widely known variant of the gas turbine is the jet engine, as used by aircraft.

Air enters the intake of a jet engine and passes through its compressor. This increases the air pressure until it is about ten times higher than its original pressure. The air next enters a combustion chamber where fuel is injected and ignited. There is a rapid rise in temperature and the resulting hot gases expand suddenly. These gases escape through the turbine, which is designed to absorb just enough power to drive the compressor, and into the jet pipe. A jet of exhaust gases emerges at high speed and the reaction thrusts the aircraft forwards.

So far four basic types of heat engine have been described; reciprocating steam engines, reciprocating internal-combustion engines, steam turbines and gas turbines. These obtain their heat from one, or more, of four basic fuels: coal, oil (including petrol), gas and nuclear material. There are several other sources of heat which have been tried, to a limited extent, including the sun and chemical reactions.

Another heat engine was the hot-air engine of the mid-nineteenth century which was rather like a steam engine, but the piston was moved by hot air instead of steam. It was used with some success but it never rivalled the steam engine. Other gases were tried, and it is interesting to note that if the cycle of operations of this type of heat engine is reversed it becomes a refrigerator.

But different heat engines were not the only rivals of the steam engine: waterwheels and later water-driven turbines, both using

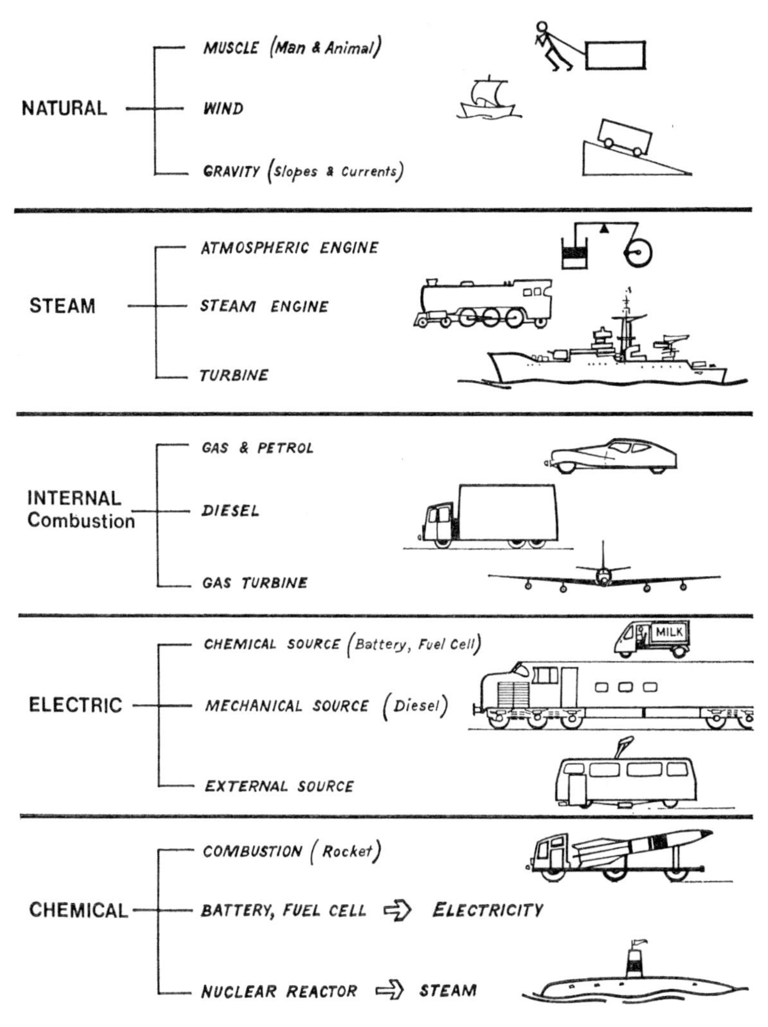

FIG 5 *Sources of power—for transport.*

natural power, have already been mentioned. Then towards the end of the nineteenth century came the electric motor which was to replace the steam engine in many fields, from trains and trams to workshops and factories.

The electric motor is a very convenient source of power because it is compact, clean and easily controlled, but it does have disadvantages. It has to be supplied with electricity, and wires are a nuisance if the motor has to move from place to place. Secondly, when electricity is produced by a generator of some kind, another form of power is necessary to drive the generator. Steam engines were used for this purpose at the end of the nineteenth century, and these produced the unusual situation of electrical power making the steam engine obsolete in some fields but creating work for it elsewhere.

During the Second World War three new heat engines were developed as practical power units. The gas turbine has already been mentioned; the other two were both developed into practical engines by German engineers. The first was called a pulse-jet, and it powered the V1 flying-bombs or 'doodlebugs'. The pulse-jet and its close relation the ram-jet are similar to a jet engine but without a compressor – and consequently without a turbine. Air is compressed as it enters the engine by the speed of the engine through the air; in other words air is 'rammed' into the intake. The ram-jet burns fuel continuously, while the pulse-jet burns it intermittently, and despite many advantages neither has fulfilled early promises.

After the flying bombs, the Germans launched rocket-powered missiles which reached a height of sixty miles before descending on their target. Rocket motors are an unusual form of heat engine because they do not require air, whereas ordinary fires need the oxygen in air to support combustion. Rocket motors are basically very simple; a highly inflammable fuel is burnt in conjunction with a substance containing oxygen, perhaps even liquid oxygen. The hot gases escape through a nozzle and the rocket lifts into the air.

Because rockets carry their own supply of oxygen they can operate in outer space where, of course, there is no air. The space-age is upon us and the rocket motor has made it possible. It is a far cry from a wheezing and puffing steam engine to a 364-foot-high space rocket, but this is the range of the heat engine.

CHAPTER 2

Early Thoughts and Experiments

Pre-1700

At the beginning of this century steam turbines, internal-combustion engines and electric motors were new and exciting inventions beginning to oust the reciprocating steam engine from its position of supremacy. Throughout the preceding two centuries the steam engine had provided power for both transport and industry. These steam engines were reliable and powerful as well as very impressive, but their size and weight were a great handicap. Steam-powered aeroplanes, washing-machines, lawn-mowers and food-mixers would not have been very practical.

Newcomen's first practical steam engine of 1712 has already been mentioned, but was this the first steam engine to be invented? The answer to this question depends on the definition of the words 'steam engine' and 'invention'. Even the apparently straightforward word 'invention' has led to many arguments and misunderstandings. For example, James Watt is often described as 'the inventor of the steam engine', and George Stephenson as the 'inventor of the railway locomotive', yet Newcomen's steam engines were widely used sixty years before Watt's, and over forty locomotives had been built before Stephenson's famous 'Rocket'.

Very few important inventions were the work of one man. Usually a number of men each made a contribution rather like the links in a chain. To add to the general confusion of who invented what, there were sometimes two chains. In other words, two men working completely independently of each other sometimes invented the same thing.

Sifting facts from fiction has given the historians many hours of research studying books, manuscripts, letters, patents and even wills. But many of the mysteries remain unsolved, shrouded by extravagant claims, counter-claims and missing evidence. Even as long ago as 1831, Stuart in his *Descriptive History of the Steam Engine* said:

'It has been remarked as a curious circumstance in the history of the Steam Engine, that almost every one who made an improvement, either in its construction or application, laid claim to the exclusive merit of having invented the engine.'

Engines were usually invented in a number of stages, or links, in the chain. Often the first contribution was not directly connected with the final invention, because it consisted of general scientific research into the properties of the atmosphere, fire, steam, heat and so on. This information enabled the second stage to emerge – plans for an engine, either as a rough idea, or as detailed drawings which might even be patented.

The next stage was to put ideas into practice and build a prototype engine or a working model. Some of these worked; some did not; but all contributed in some measure towards the final engine. Even a complete failure helped to eliminate false trails, thereby keeping the designers on the correct path to success.

A reasonably reliable working engine which could usefully be employed was the fourth stage in the development story, to be followed by the final stage, which consisted of improving the design to make it a really practical and economic source of power. From this sequence of events, and remembering the earlier paragraph about Watt and Stephenson, it is clear that the men who provided the last stage – the vital link – tended to receive most of the praise and publicity.

Sometimes an extra stage was required before one of the last three could take place. This involved overcoming some technical problem caused by such things as unsuitable materials or inadequate machine tools. The development of the jet engine, to use a modern example, could not enter its final stage of invention until metals were produced which could resist the high temperatures encountered in this type of engine.

The story of the steam engine begins many centuries before Watt and Stephenson made their contributions. In the days of Ancient Greece there were no sharply defined subjects such as physics, chemistry or engineering. Scholars studied anything or everything and were classed as philosophers. Most of their work was based on theories and deductions. Practical experiments were the exception rather than the rule, but the legend of one has survived – the displacement of water carried out by Archimedes in his bath. Philosophers such as Socrates, Plato, Aristotle and Ptolemy put their great minds to work on the mysteries of the earth and universe, and sometimes they were surprisingly near the theories of today.

A very early philosopher, Thales of Miletus, who lived some 600 years B.C., suggested that water was the prime constituent (or element) of all things. This was reasonably near the truth, but the development of this theory was not so accurate. The Greeks believed that everything was made from four elements – water, earth, air and fire. This false theory was not disproved for many centuries and it led to several other false theories which in turn had to be discredited – a process which slowed down man's understanding of the scientific principles accepted today.

To explain the mysteries of fire and combustion, the early philosophers proposed an invisible substance called 'phlogiston' which was supposed to be released by fire. Unfortunately this theory explained many experiments – incorrectly of course – but no one had a better explanation. For instance, if an object was burnt in a limited supply of air it soon went out, a fact the phlogiston theory explained by saying that the air had become saturated with phlogiston. We now know that oxygen is the key to combustion, and fire is extinguished by the lack of oxygen, not the production of phlogiston.

Even the discovery of oxygen by the English chemist Joseph Priestley in 1774 did not end the phlogiston theory. He noticed that things burnt more readily in oxygen, but explained this by saying this gas contained little or no phlogiston, in other words it was 'dephlogisticated air'. Although Priestley clung to the old theory, several scientists of his day were discarding it. One of these was a French chemist called Antoine Lavoisier, who explained the chemical composition of air and why things burn. While his work finally ended the 'phlogiston' theory of combustion, it introduced yet another false theory, this time to explain heat.

Lavoisier knew that heat was weightless, but he thought it was an invisible fluid which flowed in and out of substances, thereby changing their temperature. This fluid he called 'caloric'. Again the false theory explained many known facts and it remained in use until about 1850 – well over 100 years after the first steam engines. One conclusion which might be drawn from this fact is that the practical engineers achieved their results by trial and error, rather than by scientific theory and mathematical calculation. Whether this was the case or not is part of the story of the steam engine.

One of the first steam engines was produced during the first century A.D. by a great mathematician and inventor called Hero of

Alexandria, but it is most unlikely that he understood the theory which made it work. Hero's 'aeolipile' consisted of a hollow sphere supported by horizontal bearings and beneath this was a container in which water was heated to produce steam. This steam was passed into the sphere by way of the bearings and allowed to escape through small tangential jets pointing in opposite directions. These jets turned the sphere round and round on its bearings, making a fascinating toy which Hero did not put to any practical use. Little did he realize that he had invented a steam turbine which was a perfect example of the reaction type.

Hero also described a type of heat engine which did have a practical use – to open and close the doors of a temple. A fire was lit on the altar and, as if by magic, the temple doors opened. All the working parts were hidden from view below the floor and utilized the principle that air expands when heated. As the altar was an airtight metal tank, the fire heated trapped air, causing it to expand into another sealed container filled with water. This forced some water out and into a bucket suspended on a rope. The extra weight made the bucket fall and this, by means of ropes and pulleys, opened the doors. When the fire went out the air cooled and contracted, with the result that water was syphoned out of the bucket and a counterbalance weight closed the doors. By using the expansion of hot air to do work, this device was a forerunner of the nineteenth-century hot-air engine.

Although the great philosophers and inventors of Rome, Greece and China had only toyed with the idea of heat as a source of power, they were successful in harnessing natural power. In a few isolated cases an early invention using natural power paved the way for its successor, the steam engine. For example, pumping water was one of the important tasks in all early civilizations, and one solution to this problem was the action of a piston in a cylinder – the bicycle pump principle. Until the cylinder and piston had been invented the reciprocating steam engine could not be envisaged. The origin of the simple piston-pump is not known, but it probably came from Egypt. A philosopher from Alexandria called Ctesibius is often credited with its invention some 100 years B.C. Certainly piston-pumps were invented before Hero wrote his book *Spiritalia*, for he described one which was used as a fire-extinguisher.

Evidence of who actually invented many early mechanical masterpieces is lost for ever and all the historians can do is trace the first reference to a particular invention. These are sometimes found in

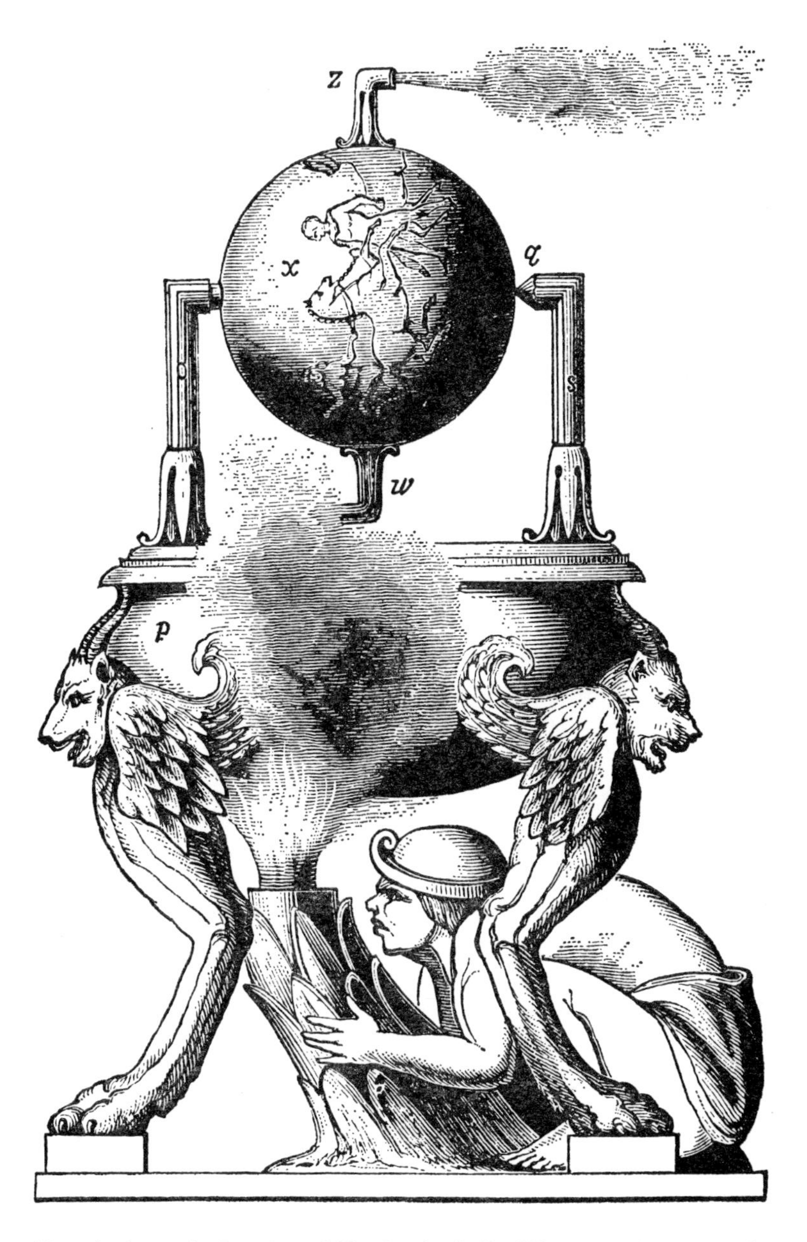

FIG 6 *An early drawing of Hero's Aeolipile. The steam jets cause the sphere to rotate—an example of a reaction type of steam turbine.*

the most unlikely places, one of these being Lake Nemi in Italy. In recent times an old barge was recovered from this lake, and archaeologists discovered that it was a Roman Emperor's ceremonial barge dating from about A.D. 40. This was an important find in itself, but in this barge were mechanisms which could be claimed to be examples of a crank and a flywheel. Although this evidence does not reveal the inventor of the crank, it does suggest that it was invented before A.D. 40.

When the Roman Empire collapsed between A.D. 400 and 500 Europe sank into the Dark Ages and little or no progress was made for about a thousand years. The emergence from this period became known as the Renaissance (or rebirth). During the early part of the Renaissance, before A.D. 1500, great progress was made in art, architecture, literature and learning. Advances in philosophy and science came later and included the great Leonardo da Vinci, who was one of the leading inventors of all times. Without decrying da Vinci however, it must be remembered that inventions pass through several stages (five have been suggested earlier) and many of da Vinci's inventions reached only the second stage – a description and drawing.

During the sixteenth century there was a great increase in the number of scientists, astronomers and mathematicians while the practitioners of magic – the astrologers and alchemists – were on the decline. For centuries alchemists had searched in vain for the secret of eternal life and the 'philosopher's stone' which would turn ordinary metals into gold. Despite their impossible aim in life, the alchemists laid the foundations of modern chemistry.

Although no new sources of power emerged between 1600 and 1700 the first steps leading directly towards a steam engine were taken. These consisted of laboratory experiments demonstrating important scientific principles, for without some knowledge of the properties of steam the steam engine would not have been invented.

When the seventeenth century dawned, England and Scotland were still separate countries (they were brought together under King James in 1603). On land, roads were very poor and people walked or travelled on horseback, while goods were carried on pack horses. Horse-drawn coaches were used but these were few in number and privately owned by the rich. In the year 1605 hackney carriages, which could be hired, were introduced in London and five years later a coach service was introduced between Edinburgh and its port, Leith. This was probably the first stage coach to run in

Britain. For short journeys in towns, sedan chairs carried by two men could be hired – but they were very expensive.

The only alternative to the horse for a long journey was to use the power of the wind. By 1600 sailing ships had carried mariners around the world on voyages of exploration and discovery. Passengers and cargo were carried to many parts of the world. On land windmills and waterwheels were replacing the horse and man as the chief sources of power for industry.

During 1606 Giovanni Battista della Porta of Naples described two laboratory experiments which were the first stages towards the invention of not one, but two types of steam engine. Of course della Porta did not realize the importance of his experiments, which were very simple. One demonstrated the power of steam, under pressure, to move water by passing steam into a sealed water tank. The pressure built up and water was forced out – rather like Hero's door-opening device but using steam in place of air. Della Porta's experiment

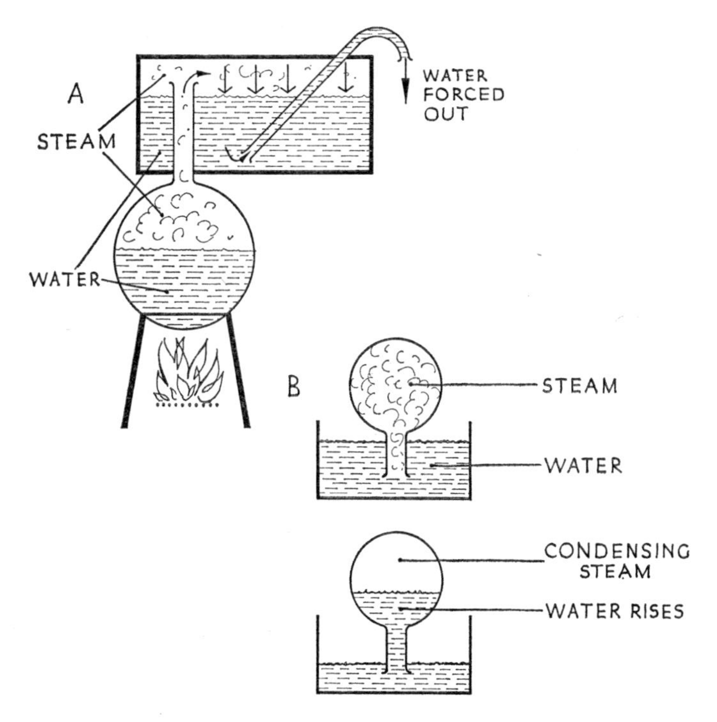

FIG 7 *Della Porta's experiments with steam.* A *Steam pressure forcing water out of a tank.* B *Condensing steam sucking water into a flask.*

demonstrated the principle of the simple steam engine, but in an engine the steam pressure moves a sliding piston in a cylinder instead of water out of a tank.

In della Porta's second experiment a flask with a long slender neck was filled with steam and held inverted, with its neck immersed in a bowl of water. As the flask cooled down, the water level inside rose well above the level in the bowl. To explain this, consider what happens when water is boiled in a kettle. Because the steam requires more space than the original water, it has to escape and in so doing it lifts the lid of the kettle or blows a whistle. Incidentally, important though this simple experiment may be in the history of the steam engine, it did not cause James Watt to invent the steam engine – this is just one example of 'fairy-tale' history which still survives. More important at this stage is the reverse of the boiling process, by which steam, when cooled, condenses back into water. If a sealed container is filled with steam and then cooled, the steam condenses into water which occupies only a tiny fraction of the space with the result that a vacuum is created inside the container. Because of the atmospheric, or air, pressure outside the container it will probably collapse, but in della Porta's experiment the flask was only sealed by water; thus the atmospheric pressure forced water into the flask. Once again the experiment only moved water and not a piston in a cylinder, but it demonstrated the principle of a second type of steam engine. Della Porta's first experiment used the power of steam under pressure: the second used the pressure of the atmosphere in conjunction with a vacuum and paved the way for the earliest steam engines called 'atmospheric engines'.

Soon after della Porta's experiment demonstrating steam pressure, a Frenchman who lived in England devised an ingenious use for steam pressure. Salomon de Caus, a notable landscape gardener, made a steam-powered fountain. A spherical water tank was placed over a fire, thus making a simple boiler. Once steam-pressure built up, a jet of water emerged from a vertical pipe through the top of the boiler. To produce a fountain it was essential that this pipe extended down into the water; otherwise steam would have emerged from the pipe – an alternative device described in 1629 by Giovanni Branca. This Italian inventor proposed to drive a wheel, rather like a horizontal waterwheel, with a jet of steam impinging on its blades. Whether or not he actually made a working model is not known, but Branca had suggested an impulse turbine in contrast to Hero's 'aeolipile' which worked on the reaction principle.

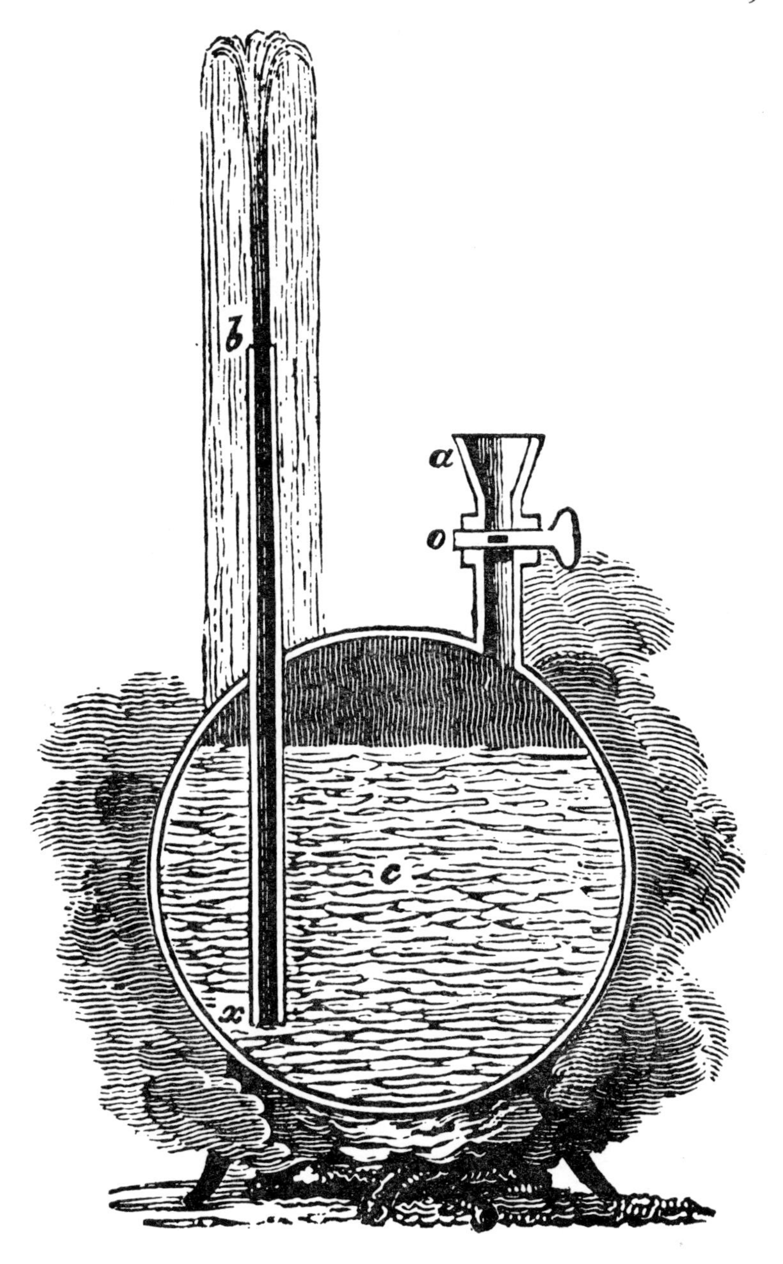

FIG 8 *De Caus' steam-powered fountain. Steam pressure forces water from the boiler* c *up the pipe* x b.

FIG 9 *Branca's steam turbine—an impulse-type. The unusual boiler* a *provides a jet of steam* b *which rotates the turbine* x.

Most of the story of steam and its harnessing as a source of power concerns scientists and engineers, but in 1624, during the reign of James I, English politicians made a contribution which was to play an important part in the story. An Act of Parliament was passed called the Statute of Monopolies, and this laid down the rules and regulations concerning patents for new inventions. Although this act was intended to protect inventors it caused many arguments and lawsuits. In 1631, a Scottish inventor patented a number of ideas of which one was 'An Apparatus to Raise Water from Lowe Pitts by Fire'. Whether the inventor, called David Ramsay, had any plans for a steam pump, or whether he was just patenting an idea in the hopes of collecting royalties from a later inventor is not known. But the dangers of such a wide-ranging patent are obvious.

Scientists of the mid-seventeenth century made many important experiments and discoveries which helped them to understand the earth and the stars. A few of these discoveries were concerned with subjects leading towards the first stage in the invention of the steam engine. While some scientists studied steam, others investigated the earth's atmosphere – particularly its pressure. Perhaps the most important work was directed towards the mysterious vacuum.

One of the greatest scientists of this period – indeed of all times – was Galileo, famous for his telescope, the pendulum and his experiments with gravity, said to have been started by dropping objects from the leaning tower of Pisa. In 1641, Galileo, then an old man of seventy-seven with a great reputation for solving scientific mysteries, was consulted by the engineers of the Grand Duke of Tuscany when they had failed in their attempt to make a suction pump. Pumps relying on suction had been made previously, but the Duke's engineers were attempting to draw water up from a depth of 50 feet – considerably further than was usually achieved. The action of this type of pump created a partial vacuum into which water was sucked, and this was explained by the statement that 'nature abhorred a vacuum'. Galileo realized that pumps would draw water up from about 28 feet and no more, a fact which he explained by saying that there was evidently a limit to nature's abhorrence. However, he started experimenting to find out more about this problem, but died the following year.

One of Galileo's pupils, called Evangelista Torricelli, continued the experiments, and in 1643 he discovered that the pressure of the atmosphere would support a column of water 32 feet high if the upper end of the tube was sealed and all the air pumped out. As this was the limit of atmospheric pressure, the attempt to draw water up from 50 feet was out of the question. A force (or pressure) pump could have been used instead. Torricelli and Viviani went a stage further and showed that the atmospheric pressure would also support a column of mercury about 30 inches high; thus with this experiment they demonstrated the simple barometer.

Progress in science was centred on the Continent of Europe at this time. Britain was in a state of turmoil due to the Civil War between the Roundheads and Cavaliers, during which King Charles I was defeated and beheaded in 1649. For eleven troubled years Britain became a Commonwealth without a king. Bitterness and revenge followed the Civil War but even as these human weaknesses were being overcome two natural disasters retarded progress; the plague killed thousands of Londoners during 1665 to be followed a year later by the great fire of London.

In the 1650s Magdeburg, a city in Germany situated on the bank of the River Elbe, had a very famous burgomaster, or mayor, called Otto von Guericke. He was not only a great scientist but also something of a showman. He carried out many experiments using a vacuum in conjunction with atmospheric pressure, and one of his first experiments demonstrated dramatically the power of this combination. Having filled a copper sphere with water, he then pumped the water out, which created such a good vacuum that the external air pressure crushed the sphere inwards with a resounding bang.

Guericke later devised an air pump to avoid using water and once again he emptied a sphere, so creating a vacuum. This time the sphere was 12 feet in diameter and suitably strengthened. It was made in two halves which were not joined together – just sealed with an airtight joint. Guericke demonstrated this experiment to the Emperor. The air was pumped out of the sphere and, in a spectacular attempt to separate the two halves, sixteen horses were used, but they could not overcome the atmospheric pressure.

A later experiment by Guericke was almost as spectacular and certainly a valuable contribution towards the steam engine. He made a cylinder with a tight-fitting sliding piston. Twenty men hauled the piston to the top of the cylinder and Guericke evacuated the air in the cylinder by connecting it to one of his vacuum spheres. The twenty men strained hard but they could not hold the piston; with atmospheric pressure on the upper side and a near vacuum below, it moved downwards with surprising ease.

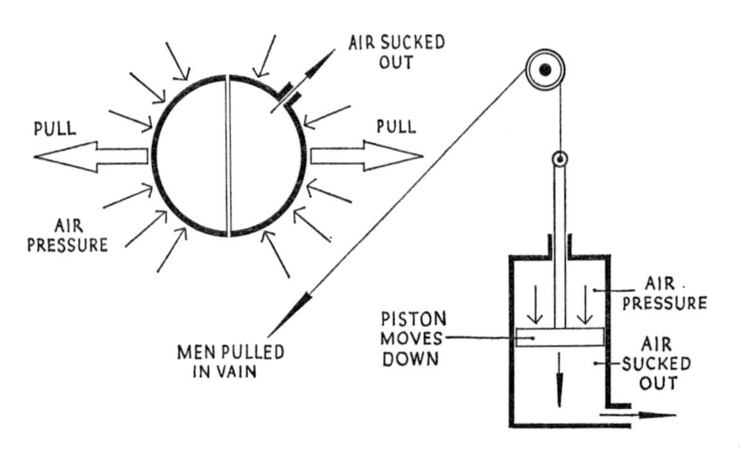

FIG 10 *Otto von Guericke's experiments demonstrating the pressure of the atmosphere.* (*Left*) *Two hemispheres.* (*Right*) *Piston in a cylinder.*

PLATE 6 *A model of a late Watt-type beam engine with a crank and a cast-iron beam—date 1840. It was double-acting with a condenser and a slide valve which was operated by an eccentric.*

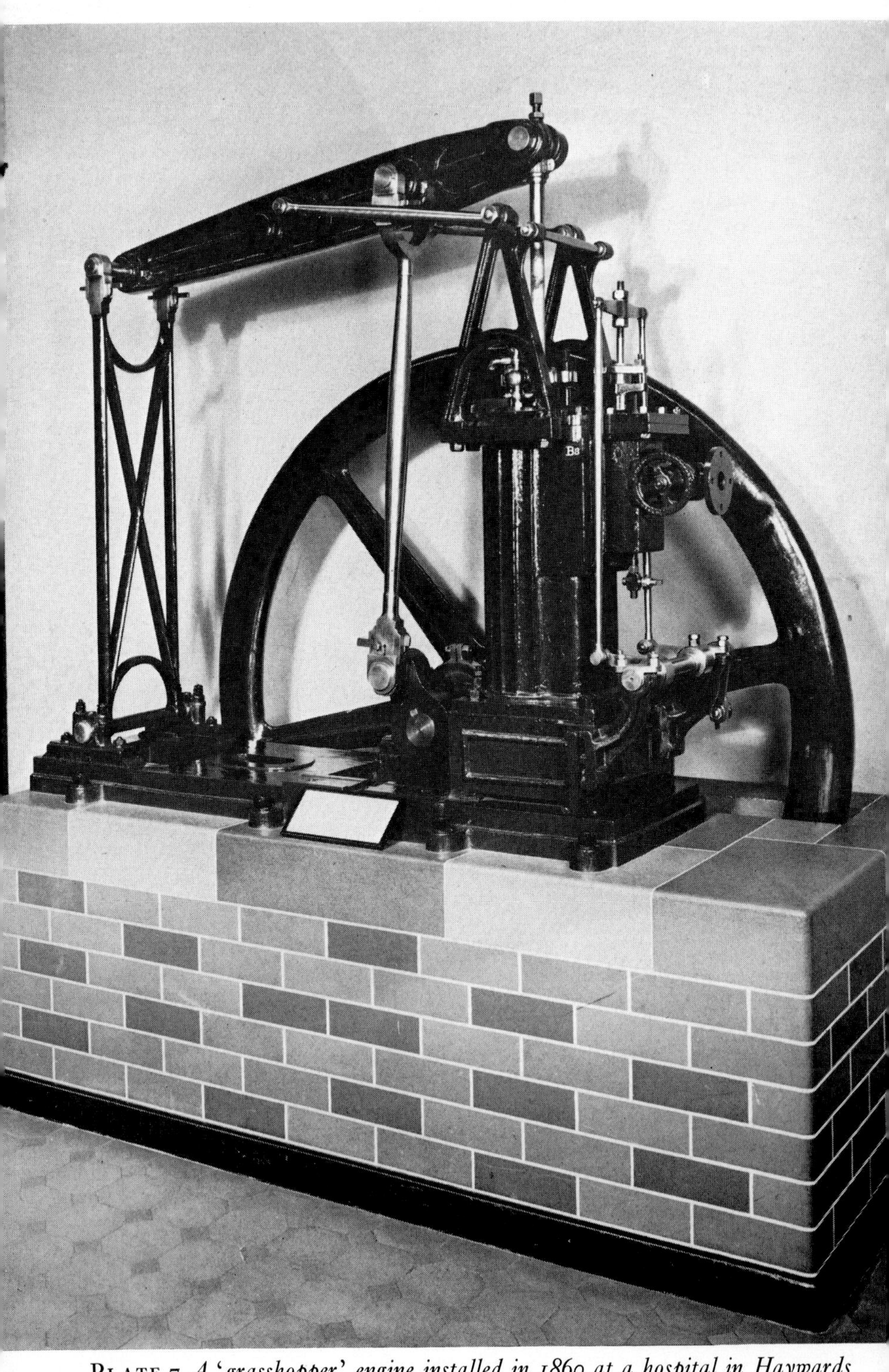

PLATE 7 *A 'grasshopper' engine installed in 1860 at a hospital in Haywards Heath. This 7½ horse-power engine drove water supply pumps until 1928 and is now in the Royal Scottish Museum, Edinburgh.*

PLATE 8 *A contemporary engraving of a Cornish pumping engine built in 1830 at New Craighall Colliery, near Edinburgh by Messrs. Claude Girdwood & Co.*

PLATE 9 *A table engine model, which is mentioned on a list of apparatus used at the University of Edinburgh in 1856.*

It was now becoming clear that a very powerful engine could be built on these lines – if only a vacuum could be produced rapidly. The air pump was not the answer. Condensing steam could produce a vacuum, as della Porta had shown by his experiment in which water was sucked upwards into a flask by the action of steam in condensing. Several inventors proposed pumping devices based on this principle, but although these were sometimes called engines, they did not have moving parts such as a piston. To avoid confusion these will be called 'steam pumps'. The Marquis of Worcester, a prolific inventor, claimed one such pump in about 1663, and Sir Samuel Morland, 'Master of Mechanicks' to Charles II, may have built one in about 1680. Unfortunately, these and other claims are almost impossible to prove one way or the other. Morland did, however, carry out a number of experiments with steam, and he published tables stating the weights which could be raised by steam pressure in various sizes of cylinders up to 6 feet in diameter.

Further theoretical and experimental work was carried out by the famous Irish chemist the Hon. Robert Boyle and his English assistant Robert Hooke. Both men have scientific laws named after them, and Boyle's law, relating the volume and pressure of a gas, was a direct contribution to the theory of heat engines known as thermodynamics. Boyle and Hooke produced an improved air pump, and in 1678 Hooke proposed a steam engine, but little is known of this device.

About the same time two separate attempts were made to produce a vacuum quickly by exploding a charge of gunpowder in a container. The air was expelled through non-return valves thereby leaving a partial vacuum inside the container. In France Jean de Hautefeuille proposed to suck water up a pipe by this method, while the great Dutch astronomer Christiaan Huygens tried to move a piston in a cylinder. Neither was very successful, but perhaps they are worthy of note because, by placing the source of heat inside the engine, they were suggesting the idea of an internal-combustion engine, which eventually led to the petrol engine.

Huygens' French assistant, Denis Papin, spent some time in England, where he worked for Robert Boyle and was proposed for the exclusive Royal Society by Robert Hooke. Papin demonstrated several of his ideas to the Society, one of which was the first pressure cooker, which he called a 'Digester'. Earlier Papin had assisted Huygens with the gunpowder experiment and he did not let the problem rest. In about 1690 Papin thought of condensing steam to produce the elusive vacuum and thereby move a piston. The power of condensing steam to suck water was not new, but the idea of using it to drive an engine was a great step forward.

Papin demonstrated his idea with a model. He used a brass cylinder $2\frac{1}{2}$ inches in diameter into which he poured a small quantity of water. A sliding piston was then placed in the cylinder and the water heated until it boiled. Steam forced the piston to the top of the cylinder where it was held by a catch. The heat was removed; the cylinder cooled down and the steam condensed. This created a partial vacuum in the cylinder, resulting in a low pressure on one side of the piston and the full atmospheric pressure on the other just like Guericke's experiment with twenty men. The catch was released and this out-of-balance pressure moved the piston downwards so effectively that it lifted a weight of 60 lb., which was attached to the piston by a string over a pulley.

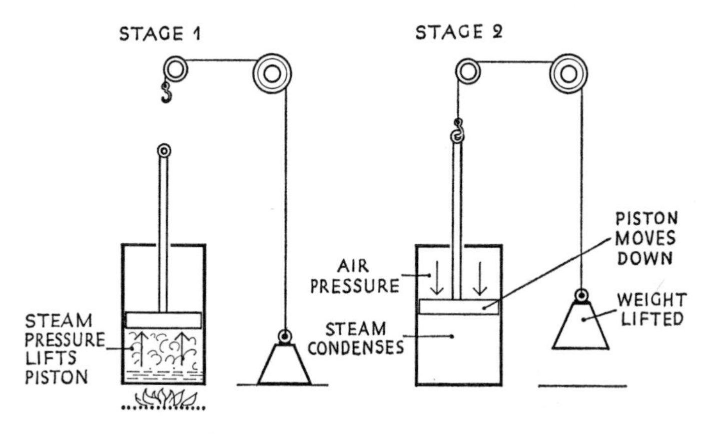

FIG II *Denis Papin's experiment using condensing steam to move a piston.*

Papin suggested larger engines with several cylinders which could be used to turn the paddles of a boat, but he did not proceed with his steam engine ideas. Perhaps he was more interested in his other activities, or perhaps the problems of manufacturing large cylinders and pistons with great accuracy deterred him – for he was a scientist, not an engineer. Papin brought the piston (or reciprocating) steam engine to its third stage – a working model.

The development of the reciprocating steam engine and the steam pump were taking place simultaneously, and each had the same alternative sources of power – steam pressure or atmospheric pressure combined with a vacuum. At the end of the seventeenth century a steam pump was built which actually used both sources of power. This was the work of Captain Thomas Savery, a Devonian inventor about whom little is known. In 1698 he patented his 'engine' for raising water '. . . by the Impellent Force of Fire', and the patent was to stay in force until 1733.

Savery was a practical man who hoped to produce a reliable pump for pumping water out of mines. The pump itself was positioned about half way up the pipe between the inlet and the outlet. Steam from a boiler was piped to a closed tank until the tank was filled with steam, then the supply was cut off and cold water poured over

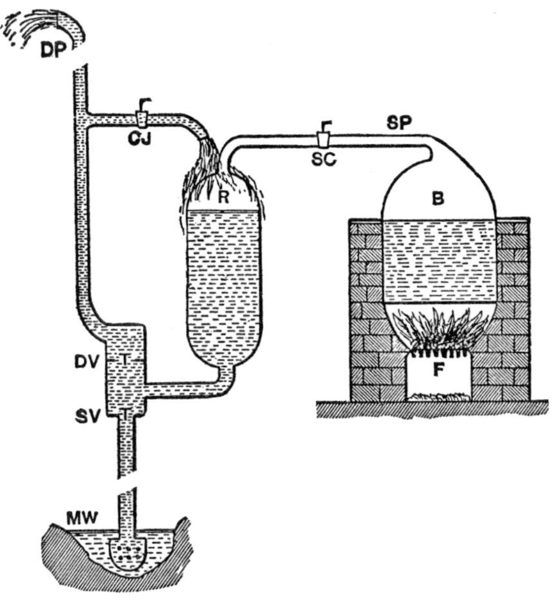

FIG 12 *A diagram of Thomas Savery's steam-powered pump of 1698.* B *boiler*, R *receiving tank*, MW *mine water*, DP *delivery pipe.*

the tank. The steam condensed, creating a vacuum, which drew water up the suction pipe into the tank (a non-return valve preventing it from escaping back down the pipe). Now the steam-cock was opened again and steam entered the tank; the pressure built up and forced the water out of the tank through a second non-return valve and up the final delivery pipe. The cycle of operations could then be repeated at frequent intervals.

There is no doubt that Savery's pumps worked, but they were little more than experimental prototypes. In 1699 he demonstrated one with twin tanks, working alternately, to the Royal Society and shortly afterwards he installed one by the Thames in London. In 1702 Savery published a book called *The Miner's Friend* in which there is a very good illustration of his pump in its final form.

The hand-operation of the valves was a slight handicap, but the chief reason for the limited success of Savery's pump was its inability to lift water more than about 50 feet. The first stage, with a perfect vacuum, should have lifted water 32 feet, but in practice 20–25 feet was the limit: the second stage, using pressure, was restricted by the boilers, which could not withstand the high pressures required. Savery's boilers were extremely dangerous as they did not have safety valves to 'let off' steam, although these had already been demonstrated by Papin on his pressure cooker.

Probably Savery's greatest contribution to the steam engine was his work improving valves and boilers, which made him one of the first great engineers to harness steam. Most of his predecessors had been scientists or just dreamers.

CHAPTER 3

The First Steam Engines

1700–1776

Events during the early years of the eighteenth century bore several similarities to corresponding years of the previous century. A new sovereign, Queen Anne, was crowned; England and Scotland were brought closer together by the union of their Parliaments, and the steam engine entered a new era. The first practical working steam engine was invented, not by a London inventor or a Continental scientist, but by an obscure ironmonger from Devon whose name was Thomas Newcomen. Despite a vast amount of research in recent years, little is known about him. No portrait of him exists and even his date of birth is unknown, but we do know that he was christened in Dartmouth on 24 February 1663.

Newcomen grew up in Devon and probably served an apprenticeship in Exeter before setting up as an ironmonger and blacksmith in Dartmouth some time around 1685. He had a partner called John Calley (or Cawley) who was a plumber, glazier and tinsmith; consequently the two men had between them a wealth of practical experience in metalwork which was to serve them well.

In 1705 Newcomen married and two years later the family moved into a large house in Dartmouth, which became a meeting-place for local Baptists. Newcomen and Calley were both devout Baptists and Newcomen was a very active preacher.

Newcomen visited the tin mines of Cornwall and Devon in the course of business and he must have been impressed with the need for a pump to remove water from these mines – especially the deep ones. How much he knew of Savery's pump or Papin's model no one knows, although historians have argued at great length on this subject. Very little is known of Newcomen's experiments and failures, for he was not in the circle of fashionable inventors, backed by wealthy patrons, who often received more than their share of publicity. Newcomen probably started his experiments about the same time as Savery, but the latter's patent prevented him from patenting his totally different idea of a water pump driven by a reciprocating steam engine. In 1705 Newcomen and Calley entered into partnership with Savery.

A Swedish engineer called Marten Triewald visited Newcomen and Calley some years later and published a book in which he mentions a model engine used for early experiments. This engine had a vertical cylinder and a sliding piston which was connected to a see-saw beam. On the other end of the beam, weights were attached to represent a mine water pump. Condensing steam was used to move the piston – as Papin had demonstrated – but Newcomen came up against a difficulty. The time taken for the cylinder to cool down and thus condense the steam was far too long for a practical engine, despite a 'jacket' or double wall fitted around the cylinder and filled with cold water. According to Triewald, Newcomen solved this problem by a fortunate accident. Water leaked into the cylinder containing steam and speeded up the condensation to such an extent that the piston moved too rapidly and the model broke. Newcomen's later engines were therefore fitted with a cold water spray inside the cylinder to speed the process of condensation.

The first full-sized engine is thought to have been erected at a Cornish tin mine near Breage, but there are no positive records of this – probably because it was a failure. Perhaps the water spray had introduced yet another problem, because it allowed small quantities of air into the cylinder. The amount of air would increase with each stroke, until the vacuum was destroyed. We know that later engines had a 'snifting' valve which was opened while steam was being admitted to the cylinder, so that incoming steam forced the unwanted air out of the cylinder through the open valve.

Presumably Newcomen could not find customers for his engines in Devon and Cornwall, but he had Baptist friends in the Midlands – a coal-mining area. Not only did the mines require mechanical water pumps, but they could also supply coal for the boiler of Newcomen's engine – a fact which may have accounted for the lack of orders in Cornwall, where coal was expensive.

In 1712 Newcomen and Calley built their first successful engine, but the Newcomen mystery prevails and there is no record of where this famous engine was installed. There is, however, a clue contained in the title of a drawing of the engine which reads 'The STEAM ENGINE near Dudley Castle. Invented by Capt. Savery, and Mr. Newcomen. Erected by ye later 1712'. Once again the detectives of history have tried to solve this mystery without success, but Tipton would appear to be the most likely site. Thomas

Barney's drawing, made in 1719, shows the engine in some detail and it includes a scale of feet and inches. In one corner there is an illustration of Dudley Castle – as seen from Tipton.

Newcomen's engine was a huge contraption occupying an engine-house some 30 feet high (about three storeys of an ordinary house), but even so the driving mechanism for the water pump was situated outside the engine-house. Dominating the layout was a 25-foot long oak beam which was pivoted at its centre and rocked like a giant see-saw up near the roof. The purpose of this beam was to transfer power from the engine's single cylinder to the water pump, and not surprisingly this layout became known as a 'beam engine'. Cumbersome though they were, beam engines remained almost unchallenged by any other form of heat engine for over a hundred years.

The working sequence of Newcomen's engine can be more easily understood if a simplified diagram is studied. Under one end of the beam – or 'Great Lever' as Barney described it – stood a long brass cylinder, and beneath this a domed boiler. The water inside the boiler was heated from underneath by a coal fire. Inside the brass cylinder a piston moved up and down, a movement which was transferred by a chain to the end of the 'see-saw' beam. From the other end of this beam, a short chain was attached to rods which extended 150 feet down into the mine to drive a water pump.

The pump end of the see-saw was deliberately made heavier to ensure that it rested in the 'down' position, so the other end, and with it the piston, started in the 'up' position. With the piston at the top of the cylinder, a steam valve between the boiler and the bottom of the cylinder was opened. The cylinder filled with steam. The valve was closed and a water injection cock opened for a short time to spray cold water into the cylinder. When the steam condensed, a vacuum was created inside the cylinder allowing the outside atmospheric pressure to force the piston downwards. At the same time a vertical rod and chain from the piston pulled the end of the beam downwards. Incidentally, to ensure that the piston rod moved in a straight line, the chain was attached to the beam by means of a curved 'arch head'. While the piston moved down on a power stroke the other end of the beam moved up. Another arch head, chain and rods transferred this power to the pump which raised about 10 gallons of water with each stroke. When the piston neared the bottom of the cylinder the working stroke came to an end, and the heavier weight at the other end of the see-saw returned the piston to the top of the cylinder ready for the next working stroke.

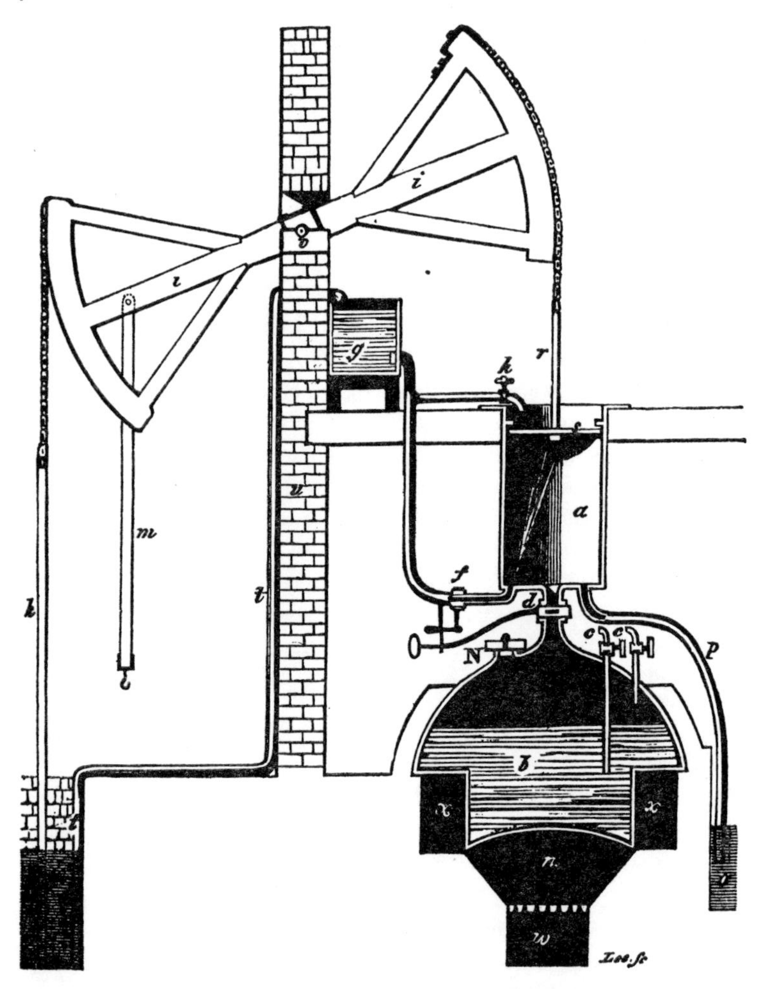

FIG 13 *A simplified diagram of a Newcomen engine showing: the boiler* b, *the cylinder* a (*note the spray of cold water to condense the steam*), *the piston* s, *the beam* i *and the rod to the pump* k.

Steam was allowed into the cylinder during the up-stroke, and this helped to drive out the water left by the spray and condensing steam. The drain was called an 'eduction' pipe and was fitted with a non-return valve which allowed water to flow out, but automatically sealed itself if air or water tried to enter. As already described, air was bled off through a snifting valve.

The cylinder had an internal diameter of 19 inches, and the vertical movement, or stroke, of the piston was about 6 feet. Working at a steady 12 strokes per minute, this first successful engine developed about $5\frac{1}{2}$ horse power.

Newcomen showed great ingenuity in the detailed design of his engine, for, although several of these details were not new, they had to be adapted for the steam engine. One major problem was making the piston a leakproof fit inside a cylinder which had to be shaped by hand. The lower face of the piston was made as close a fit as possible, while the upper part was considerably smaller. Into this annular recess, soft hemp packing was coiled round and round until the space was filled, then weights were placed on top of the packing to force it against the cylinder walls. In order to maintain a good seal the hemp had to be kept wet, and this was achieved by pouring water on to the upper side of the piston. A tank to supply water for the piston seal and the spray inside the cylinder was placed high above the ground, near the beam. Water was pumped into this tank by a ground-level pump which was driven by a long vertical rod attached to the beam. The up-and-down movement of this 'plug rod' was also used to open and close the steam and water injection valves.

Barney's drawing of 1719 poses another problem, for it illustrates an engine in which the valves were opened and closed automatically by the plug rod. The valves and steam cocks on Newcomen's early engines were often described as being operated by hand. Legend has it that a lazy boy devised a system of rods and string connected to the rocking beam which did his work for him. This story is difficult to believe because the complete mechanism is rather complicated for a boy to invent. Newcomen probably developed automatically operated valves before 1712, but his engines could be speeded up when their boilers were making plenty of steam by 'tripping' the valve mechanism manually. This simple operation undoubtedly was carried out by boys and it could easily have been made automatic by an intelligent but lazy boy. In fact, the simple mechanism was later called a 'Potter cord', reputedly after a boy named Humphry Potter.

Before proceeding with details of other engines it might be advisable to summarize the various descriptive names. 'Beam engine' describes the basic layout which could incorporate one of two sources of power – direct steam pressure, or atmospheric pressure used in conjunction with condensing steam. The latter group are called 'atmospheric' engines, and this group can again be subdivided into two according to the method of condensing steam. If this took place within the cylinder it is called a 'Newcomen-type' engine. (The alternative method of condensing is still in the future at this stage in the story.) Later engineers made some improvements to Newcomen's design, but the basic essentials remained the same and it was still being used for engines built a hundred years later.

When Savery died in 1715 the patent passed into the hands of a London syndicate who kept Newcomen in the background and restricted the number of engines being built by charging very high royalty fees. Unfortunately Newcomen did not benefit financially from his invention during this period, nor did he have the satisfaction of seeing the rapid increase in popularity of his engine when the patent expired in 1733. He died almost unknown in 1729.

One of the engineers who contributed to the early success of the atmospheric engine was Henry Beighton of Warwickshire, a well-educated young man, aged 26 when Newcomen built the first engine. Beighton devised an alternative system for operating valves, but perhaps of more importance was the fact that he studied the performance of steam engines. In 1721 he published a table showing the size of engine needed to pump various quantities of water. This useful information appeared in a most unlikely publication, edited by Beighton – *The Ladies' Diary*. The versatile Beighton also left a drawing and a map. His drawing, made as early as 1717, shows an atmospheric engine, while his map of Warwickshire, published in 1725, marks three 'fire-engines' (a name often used to describe these early steam engines).

By 1725 the atmospheric steam engine was well established, but little progress had been made with the alternative source of steam power – utilizing high pressures. One reason for this bias towards the atmospheric engine was the absence of a reliable boiler. It was enough of a problem to raise large quantities of steam with the primitive and leaking boilers, without expecting them to withstand pressures of, say, 50 pounds per square inch.

The idea of using high-pressure steam was not overlooked, for in 1725 Jacob Leupold of Leipzig designed a steam engine on this principle, which also utilized two cylinders. High-pressure steam was to be supplied to each cylinder in turn, and after moving the piston the steam would be released into the atmosphere. It is not

FIG 14 (*Opposite*) *Leupold's high-pressure engine of 1725, with two cylinders* r *and* s *powering pumps* o *and* p.

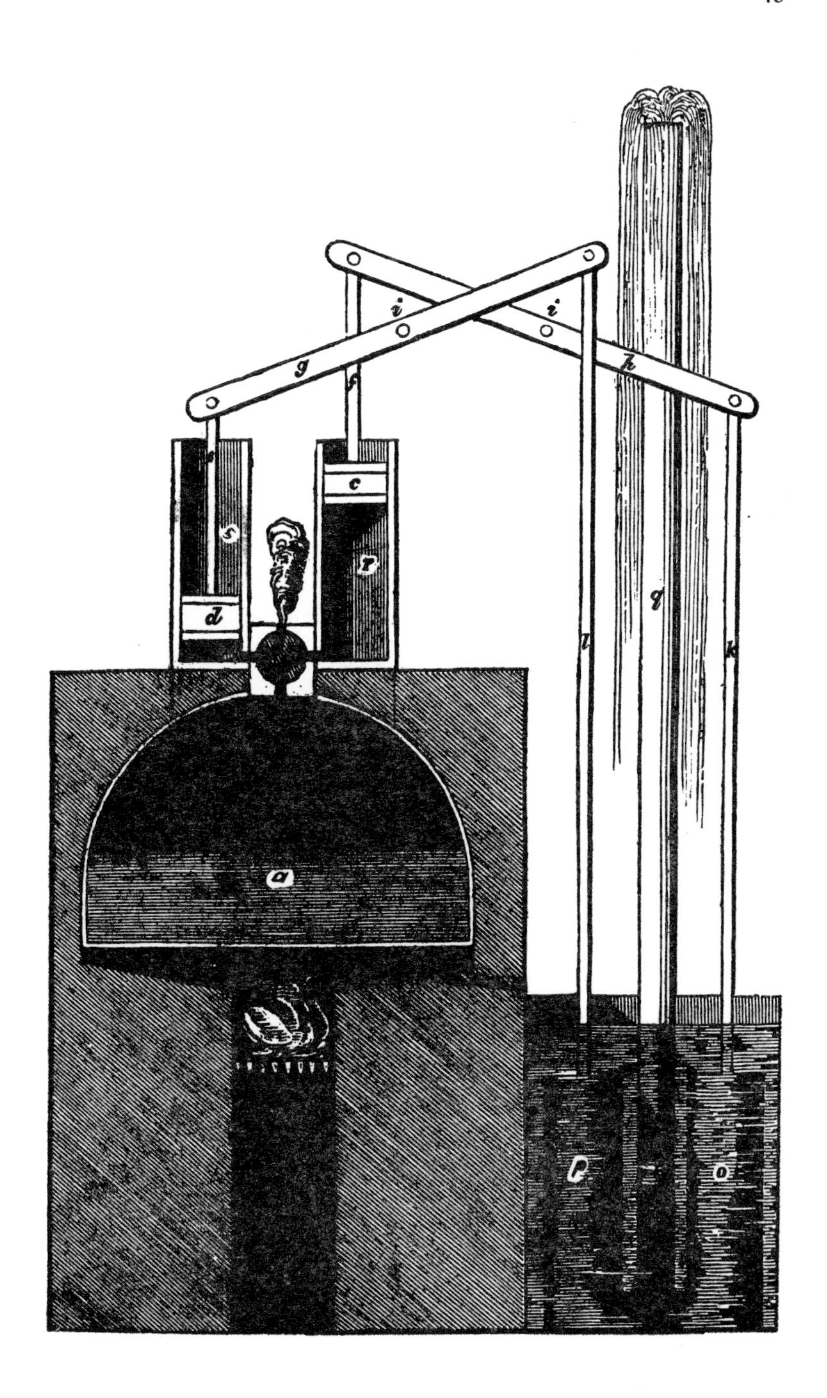
i
i'
g
h
f
c
s
r
d
q
l
k
a
p
o

known whether this pumping engine was ever built, but it is interesting to note that Stuart reports in his book of 1831:

> With a candour unusual in the history of the Steam Engine, Leupold ascribes the sole merit of his contrivance to Dr. Papin, as it was to him, he confesses, that he was indebted for the idea of employing the elastic force of steam to raise water.

It was to be some seventy-five years before the practical difficulties of using high-pressure steam could be overcome, but in the meantime the Newcomen-type of atmospheric engine improved steadily. The speed of progress was often limited by the materials and shaping machinery available at the time. For instance, the early cylinders were made in brass and largely hand-shaped, both of these factors making them expensive. In 1709, however, Abraham Darby founded his Coalbrookdale Works in Shropshire and started to produce cast iron by a new and improved process. He burnt coke instead of the traditional charcoal to smelt iron ore, which resulted in better-quality iron at a lower price. Cast iron became considerably cheaper than brass, and it is not surprising to find that cast-iron steam engine cylinders were being made at Coalbrookdale. Company records show that twenty-seven were made between 1722 and 1733.

In 1733, Savery's patent, held by the London syndicate, expired and engine owners no longer had to pay royalties. The Coalbrookdale Company prepared to manufacture more cylinders, and to do this they had to overcome the problem of boring the large hole down the middle of each cylinder. This hole, between two and three feet in diameter, was machined on a boring mill which used a long boring bar to cut away the metal inside the cylinder. These boring bars frequently broke. In 1734 a large new wrought-iron bar was ordered for Coalbrookdale, which had a length of 12 feet and a diameter of 3 inches. This must have eased the problem, for a second one was ordered in 1745.

1745 was also the year of the rebellion by the Scottish Highlanders under 'Bonny Prince Charlie' which ended the following year at the Battle of Culloden – the last battle fought on British soil. About the same time a political battle was won which started a revolution in the steam engine business. For many years the Government had been requested to reduce their tax on coal because it was crippling the engine owners. Large quantities of coal had to be burnt in order to raise steam in the rather inefficient boilers, and even the end of royalty payments in 1733 did not result in the expected increase in the number of steam engines. Eight years later when the Government eased the taxation the numbers grew steadily, particularly in the coalfields of the North and Midlands and in the tin-mining area of Cornwall.

Almost all the steam engines built at this time were used for pumping water out of mines, and as the mines grew larger and deeper the problem of pumping became more difficult. Some mines used as many as six pumping engines to obtain sufficient power, but another solution became available as manufacturing methods improved – that of larger engines. In 1745 cylinders with a bore of 36 inches were considered large, but less than twenty years later engines were being built with cylinders of more than twice this diameter. These engines must have been an impressive sight with a piston, six feet in diameter, pounding and hissing up and down every six seconds. Three or four boilers were needed to supply sufficient steam, and these would burn more than 5 tons of coal every day.

On a June day in 1753 a ship sailed from London with New York as her destination. In itself this was not a very unusual event, but she had an important passenger and an unusual item of cargo. The passenger was Josiah Hornblower and he was taking the first steam engine to America. Josiah's father had worked with Thomas Newcomen, and his brother Jonathan was a famous engine-builder in Cornwall. Josiah erected the atmospheric engine at a copper mine in New Jersey and settled in America. After the rough Atlantic crossing he is reported to have vowed never to go to sea again!

During the latter part of the eighteenth century several great engineers emerged and, unlike their modern counterparts, they became proficient in several branches of engineering. James Brindley, who became famous for his canals, also built a steam engine in Staffordshire during 1756. Another famous all-round engineer was John Smeaton, who is perhaps best known for his Eddystone Lighthouse. Two earlier lighthouses had been built of wood, but Smeaton built one of stone which withstood the battering waves from 1759 until 1886, when it was removed to Plymouth Hoe where it can still be seen. Smeaton, as well as building bridges, waterworks, canals and his own observatory, also played an important part in the history of the steam engine, not by any revolutionary invention but by improving the details of Newcomen's design. Other engineers had made some improvements, but these were usually the result of 'trial and error', whereas Smeaton studied the 'whys and wherefores'.

In 1765 Smeaton built a model engine to gain information before building a full-size engine at Islington to pump water into a reservoir. When the engine was built two years later it was not as efficient as Smeaton had hoped; nevertheless he had discovered an important phenomenon called 'scale-effect'. Model tests were a useful guide, but scaling-up their results to full-size engines led to serious errors. For example, doubling the dimensions of a cylinder increases the area of the piston four times and the volume of the cylinder eight times. Smeaton realized that the only way to improve his engines was to study existing working engines and compare their performances. Several steam engine builders helped Smeaton with a survey which they carried out in about 1769.

The survey revealed that many of the engines were not only badly made but also operated in an inefficient manner. Smeaton wanted to be more specific about efficiency, and to do this he worked out a measure of efficiency which was called the 'Duty'. As all the engines were used for pumping water, the duty was a measure of the quantity of water which could be raised one foot for every bushel (84 lb.) of coal burnt in the boiler. Since the resulting figure was large, the water was measured in millions of pounds. Newcomen's early engines returned a duty of just over 4 (i.e. 4 million foot-pounds per bushel) while the best in Smeaton's survey recorded 7·44. After his studies Smeaton built an improved engine at Long Benton Colliery in Northumberland in 1772 and this gave a duty of 9·5.

Although the duty was a useful measure of the efficiency of an engine it gave no indication of power. Smeaton therefore calculated the quantity of water which could be raised one foot in one minute and called the figure the 'Great Product'. This is closely related to horse-power, but in Smeaton's day the unit of horse-power had not been defined. The Long Benton engine developed 40 horse-power compared with Newcomen's first engine which produced some 5 or 6 horse-power.

The Long Benton engine had a cylinder 52 inches in diameter, and Smeaton followed this with a 72-inch engine in Cornwall which developed about 75 h.p., and a 66-inch engine in Russia. The latter was used to pump water out of the dry docks at Kronstadt, a task it could complete in two weeks in contrast to the period of one year required by numerous windmills.

Parts for these engines were made at the Carron ironworks in Scotland. This company began production in 1760 and rapidly challenged Coalbrookdale's supremacy. Smeaton assisted by designing some of their equipment, including a new cylinder-boring machine and a pump to blow air into the furnaces, both of which were powered by waterwheels.

During the 1760s several other events took place which were important in the story of the steam engine. Perhaps the least-noticed event at the time, but the one with the most important long-term consequences, took place in 1764, when a model of a Newcomen-type engine was brought for repair to James Watt, at that time instrument-maker to Glasgow University. The vital part played by Watt in the development of the steam engine is the subject of the following chapter.

Improved versions of Savery's pump were suggested by several inventors including William Blakey of London in 1766, but these had little practical success. The principle was sound – a fact which was proved in the late nineteenth century when eventually a pump called a 'Pulsometer' was made to work very successfully using Savery's idea.

Having discovered that the steam engine was a reliable source of power, it was natural that engineers would not be satisfied with the up-and-down motion of a pump – they would want to drive rotating mills and machinery. Of course pumping engines had been used to supply water for rotating waterwheels which, in turn, drove machinery, but the problem of a direct drive had still to be solved.

The Newcomen-type beam engine was not the ideal source of power for rotary motion. It was single-acting, which meant a power stroke in one direction only, followed by a return stroke due to gravity acting on a counterbalance weight. This produced a rather jerky movement, whereas its rival, the waterwheel, turned smoothly and was easily controlled. The pump end of the rocking beam could have been adapted to turn a shaft by means of a connecting rod with a crank, and a large flywheel would have smoothed out some of the jerky movement. But, despite the fact that the crank had been used for centuries on such things as spinning wheels, it was not adapted for steam engines at this stage. Probably the engineers were dubious about the ability of a crankshaft to control the powerful piston at each end of a stroke.

Several inventors designed clever ratchet devices which drove a shaft round when the piston moved in one direction, but clicked free on the return stroke – rather like the freewheel ratchet on a bicycle. One of the first attempts to use this type of mechanism was at a colliery in Northumberland during 1763. A Newcomen-type

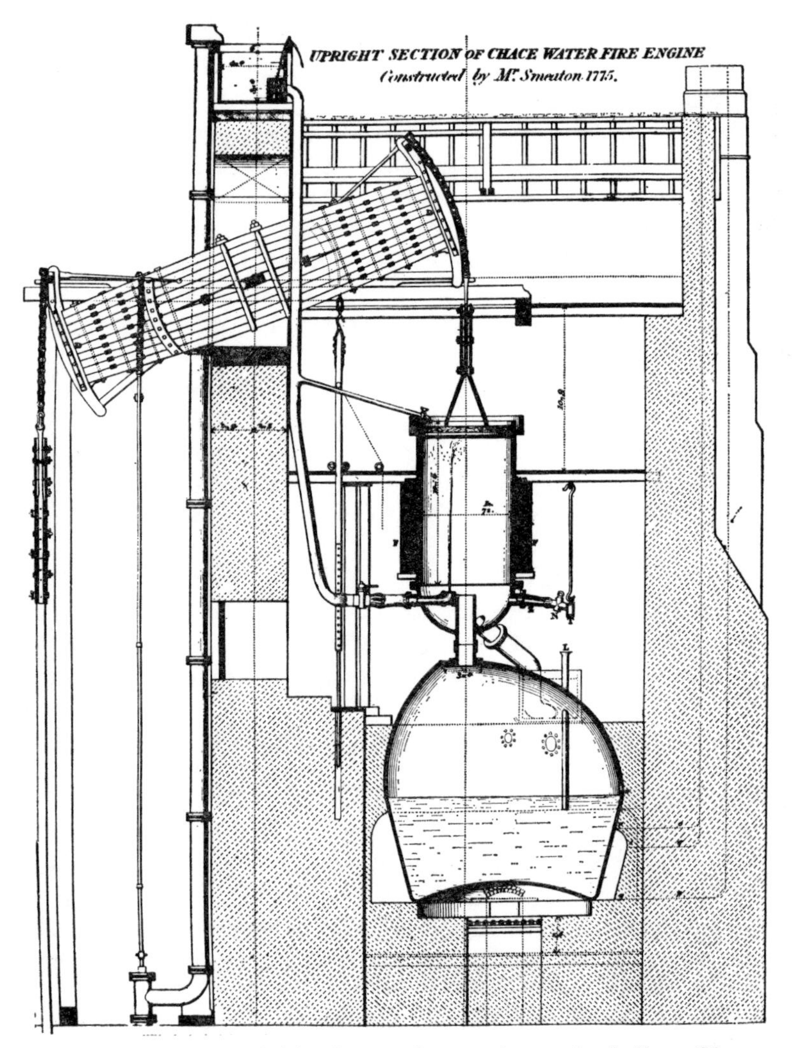

FIG 15 *A drawing of John Smeaton's pumping engine built at Chacewater, Cornwall, in 1775.*

engine was adapted by John Oxley to turn a shaft and raise coal from the mine. Its irregular rotation resulted in the engine being replaced by a waterwheel, and the engine being relegated to pumping water. Several other attempts were made, without any real success.

A very interesting machine using a ratchet drive was produced in France by a military engineer called Nicholas Cugnot. He designed and built the first mechanically propelled road vehicle in 1769. Powered by a steam engine, it trundled along at a slow walking pace – stopping every fifteen minutes to pick up water. Cugnot hoped that his *fardier* would be used by the army to haul cannon or ammunition, and in 1770 he gave a demonstration before military experts. It performed reasonably well and carried a number of passengers. Although it crashed later, the Government were sufficiently impressed to order a second and larger version which appeared in 1771. Unfortunately there was a change in Government and the project was abandoned – teams of horses proving to be more popular and versatile. The second *fardier* was later preserved in a Paris Museum, where it can still be seen.

The layout of the *fardier* was rather unusual, for it had three wheels, with the engine and boiler mounted above and forward of the single, steerable front wheel. All this weight on the steerable wheel naturally made the vehicle very clumsy and difficult to handle, and in addition it had to stop when the boiler needed stoking. Nevertheless, it did work.

Cugnot's steam engine was interesting because it was not a Newcomen-type of atmospheric engine, instead it used high-pressure steam. It also employed two cylinders, with one piston connected to each end of a see-saw beam. Steam was fed into each cylinder in turn, and the downward movement of the piston under a power stroke rocked the beam, which in turn lifted the idle piston to the top of the cylinder ready for its power stroke. In many ways this vehicle was the forerunner of the traction engine.

During the year of 1776 the Americans issued their Declaration of Independence, and a few years later British rule ended. The same year saw the beginning of the end of the Newcomen-type of atmospheric engine, for it was in this year that James Watt built his first successful engine with a separate condenser. The new type of engine did not immediately supersede the old, and for more than a quarter of a century both were being built. In addition, some Newcomen-types were modified with Watt's improvements – a fact which has resulted in very few of the former being preserved in their original state. Most of the Newcomen-type engines which have survived have done so because they were still being used at the beginning of the present century. It was natural that, in a working life of 125 years, improvements were made and parts replaced. For example, the wooden beam made in the eighteenth century was usually replaced in the nineteenth century by one made in cast iron.

CHAPTER 4

James Watt

And the improved steam engine

In the year 1760 King George II died, to be succeeded by his grandson, who became George III or 'Farmer George'. By a strange coincidence this monarch, who was passionately fond of farming, came to the throne at a time when Britain was becoming the world's leading industrial nation. The change has become known as the Industrial Revolution and it lasted about a hundred years, for sixty of which 'Farmer George' was King.

The steam engine played an important part in this revolution – but not in the early years. There were two reasons for the delayed impact; one, already mentioned, was the inability of the existing steam engines to do anything but pump water; secondly, until factories had been established there was no real demand for mechanical power. In the days before 1760 most industries were carried out in the home or in small workshops where the muscles of men, and women, provided the power. When the change came it was a gradual process. A good example of a home industry was the spinning and weaving of cloth.

In about 1764 James Hargreaves made his famous 'Spinning Jenny' which could spin yarn at least eight times as fast as the old spinning wheel. A further improvement came a few years later when another type of spinning machine was constructed by Richard Arkwright. This proved to be rather hard to turn by hand, with the result that Arkwright drove it by means of a waterwheel. By combining the ideas of Hargreaves and Arkwright, Samuel Crompton produced a hybrid spinning machine appropriately named a 'Mule'. His first machine, of 1779, had forty-eight spindles producing yarn, but an important feature was its ability to spin a very fine and strong yarn. All these improvements in spinning left the weavers struggling to cope with the output of yarn, and in 1785 a new era in weaving began. A patent for a powered loom was taken out by Edmund Cartwright – a parson, poet and inventor. A few years later Cartwright opened a mill in Doncaster equipped with powered looms and spinning machines. The power included a steam engine with a cylinder 42 inches in diameter.

Thus, a quarter of a century after the spinning wheel started to become obsolete, a steam-powered factory had been established. Unfortunately it was a financial disaster, due largely to Cartwright's inexperience in business and manufacturing methods. Nevertheless, powered machines grouped in factories replaced home industries and the use of waterwheels spread rapidly. Such factories had, naturally, to be situated near rivers, which was not always convenient, and the need for steam power grew.

The Industrial Revolution was not confined to the factories: it also spread into the field of transport. As more and more factories were built it soon became apparent that road transport could not cope with the vast increase in the traffic of raw materials, machinery and finished products. Because of the poor surfaces on the roads, heavy and bulky objects were carried by sea or river whenever possible. The success of water transport led to the construction of artificial inland waterways. In 1761 James Brindley completed his famous canal for the Duke of Bridgewater to carry coal from the Duke's mines seven miles into Manchester. It proved to be a great success, as operating barges was so much cheaper than pack-horses and the price of coal in Manchester was halved.

A great network of canals was built during the Industrial Revolution, linking London and Bristol with the manufacturing areas of Yorkshire and Lancashire. Incidentally, as the word 'Navigation' was used to describe all forms of water transport, including the canals, the workers who dug the canals were called 'navigators'. This was shortened to 'navvies' – a word which became part of the English language – and its meaning was extended to include the tough manual workers who built the railways and roads.

The canals provided transport for goods, but passengers still travelled by road. Even as late as 1775 there were still only about 400 stage coaches on the roads of Britain. In this year James Watt travelled to London – presumably by stage coach – in order to arrange an extension to his patent for an improved steam engine. His original patent of 1769 had only a few years to run, and so far the only engine he had built was an experimental one. However, by 1775 Watt and his partner Matthew Boulton were ready to sell the engine commercially, and it was for this purpose that Watt was journeying to London.

Contrary to the popular story, James Watt did not obtain his inspiration by watching a kettle lid lift due to a build-up of steam pressure: in any case his engines were of the 'atmospheric' type. While repairing a model of a Newcomen-type engine for Glasgow University, Watt realized that a large quantity of heat was being wasted by successively heating the cylinder to fill it with steam and cooling it to condense the steam. Watt solved this problem, not by sitting in front of a fire but in his own words:—

> 'It was in the Green of Glasgow. I had gone to take a walk on a fine Sabbath afternoon. I had entered the Green by a gate at the foot of Charlotte Street – had passed the old washing-house. I was thinking upon the engine at the time and had gone as far as the Herd's house when the idea came into my mind, that as steam was an elastic body it would rush into a vacuum, and if a communication was made between cylinder and an exhausted vessel, it would rush into it, and might be there condensed without cooling the cylinder. . . . I had not walked further than the Golf-house when the whole thing was arranged in my mind.'

It being a Sunday, Watt restrained his obvious excitement until the following day, when he started work on a model to test his idea. Although it did not work particularly well it was enough to encourage Watt to further experiments, and this historic little model is preserved in the Science Museum, South Kensington. Like all steam engines, it had a cylinder and sliding piston, but, in addition, it was connected to two other important components, which were a condenser and an air pump. Watt avoided cooling his cylinder by condensing the steam in the separate condenser, which was connected to the cylinder. This condenser was kept cool with cold water while the cylinder remained hot – it even had two walls between which steam passed. The key to the success of this operation was the air pump which, by sucking air out of the condenser, created a partial vacuum. Steam from the cylinder rushed into this cold vacuum and immediately condensed. The surrounding steam at atmospheric pressure then moved the piston.

The date of Watt's experiments with the model was May 1765, but ten years were to pass before a practical engine was ready for sale. Part of this time was spent experimenting with a full-size engine which was erected at Kinneil House, near Stirling – the home of Dr John Roebuck, founder of the Carron Ironworks. Roebuck had previously used atmospheric engines and made cylinders for

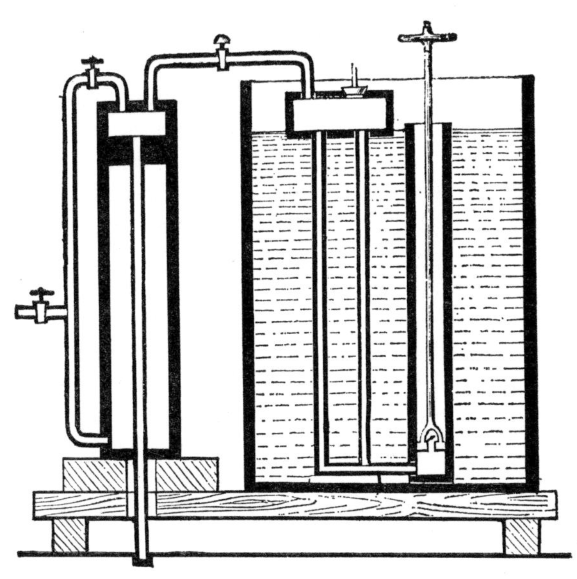

FIG 16 *Drawing of Watt's experimental model with a separate condenser, 1765. The piston and cylinder are on the left, the air pump is on the right and the condenser in the centre. This simple condenser consists of two thin tin pipes cooled by a tank of water.*

Smeaton, but he was impressed by Watt's new ideas and from 1768 he financed the building of a full-size engine.

Watt obtained his patent in 1769, but progress was slow for several reasons – one being Watt's tendency to drift from one project to another. He was a maker of scientific instruments by trade and ran a shop, but he also repaired musical instruments, had an interest in a pottery and became a surveyor. During any spare time his fertile mind was striving to invent all manner of things. Without delving too deeply into Watt's interesting life, it is obvious that he needed a staunch friend and adviser to channel his genius towards a useful end. In 1773 Roebuck got into financial difficulties, and Matthew Boulton acquired his share in Watt's patent and a very successful partnership began.

Boulton, who was a great organizer and shrewd businessman, had a factory in the Soho district of Birmingham where he employed about 600 highly skilled craftsmen. Watt was delighted with the precision work produced in the Soho factory because, as an instrument-maker, he appreciated fine work, but more important, his steam engine parts had to be made very accurately in order to work efficiently.

Watt moved from Glasgow to Birmingham and took with him his experimental engine. This was still giving trouble – the chief difficulty being leakage between the piston and the uneven walls of the cylinder. Boulton suggested ordering a new cylinder from John Wilkinson, who described himself as an 'ironmaster'. The Darby family at Coalbrookdale had played an important part in the success of Newcomen's engine by casting and machining cylinders, but Watt demanded higher standards which could be provided only by the rival Wilkinson family at Wrexham. John Wilkinson had patented a new machine for boring guns and cannon. By means of improved cutting tools Wilkinson's machine could produce a bore which was both circular and parallel, whereas earlier boring machines, such as Smeaton's, could produce a circular cylinder but with sides which were often far from parallel.

A new cast iron cylinder from Wilkinson's foundry arrived at Soho in the spring of 1775 and this was fitted with an improved piston. Watt had previously departed from Newcomen's design of piston which used a hemp (i.e. rope fibre) seal, and tried to obtain an airtight seal with a variety of substances packed around the piston, including horse and cow dung, papier mâché, felt and pasteboard. Watt eventually returned to a seal made from hemp and oakum (i.e. old rope fibre), but his design differed from Newcomen's in as much as it was packed with fatty tallow instead of using water to keep the packing soft. This troublesome gap between the piston and cylinder wall was later – and still is – sealed by springy metal piston rings.

Having overcome the problem of sealing the piston, Watt's experimental engine worked sufficiently well for Boulton to persuade Watt that the time had come to sell some engines. Orders for two beam engines were received during the summer of 1775, one from John Wilkinson and the other from a colliery near Tipton. By one of those strange coincidences Tipton was also the probable site of Newcomen's first engine. Wilkinson cast and bored cylinders for both these engines; his own had a diameter of 38 inches and was installed to blow air into his blast furnaces which produced cast iron. The colliery engine was larger, having a 50-inch cylinder and was destined to pump water.

The Boulton and Watt beam engines were not unlike Newcomen-type engines in overall appearance, particularly as the condenser and airpump were below ground level. Their debut in 1776 aroused a certain amount of interest, but few people realized the full significance of the new type of engine, as this only became apparent when figures of coal consumption were available. The 38-inch Boulton and Watt engine installed for John Wilkinson replaced a 49-inch Newcomen-type which barely supplied enough air for the furnaces, despite a steam supply from two boilers. The new engine supplied more than enough air and needed only one boiler.

The separate condenser was not the only improvement introduced by Watt in his design, for he also covered over the top of his cylinder and used steam to assist the piston downwards on its power stroke. In the Newcomen design the top of the cylinder was open to the atmosphere, and air pressure moved the piston downwards when a vacuum was formed below the piston. Watt still used the vacuum but he replaced the air pressure by steam pressure, a change which

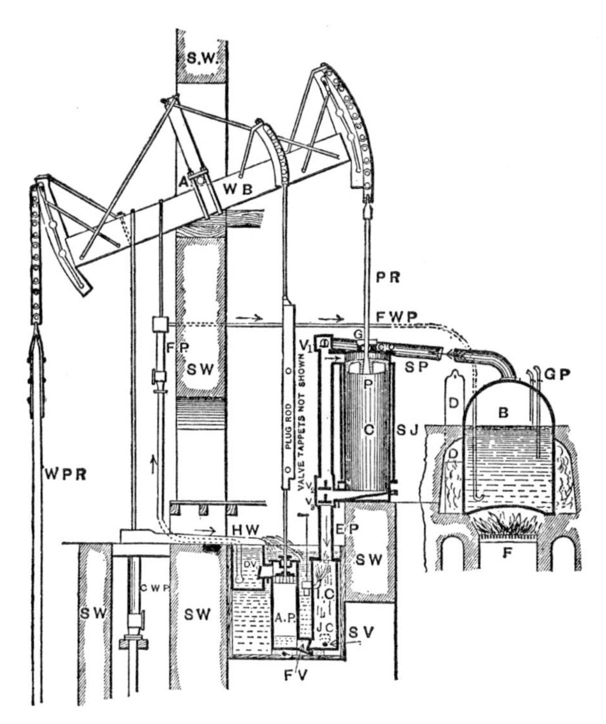

FIG 17 *James Watt's early type of beam engine used for pumping water.* WB *Wooden beam;* A.P. *Air pump;* A *Pivot of beam;* WPR *Weighted pump rods;* B *Boiler;* C *Cylinder;* P *Piston;* JC *Jet Condenser.*

had a dual advantage. It kept the piston and cylinder hotter and it gave more power because the steam was at a higher pressure than the air. Watt appreciated the power of high-pressure steam – his patent included its use – but he preferred to rely on a vacuum and low-pressure steam. To use high pressures introduced practical problems such as leaking joints and even exploding boilers. Watt was a cautious man who believed in advancing steadily, step by step.

Of course there were troubles with the new engines and they required minor modifications, but Boulton's bold decision to put them on the market paid off. If left to his own devices Watt would have carried on experimenting for years, and probably somebody else would have stepped in and commercialized his idea – a fate which all too often befell inventors. As the financial and legal aspects rarely interested an inventive mind, inventors were often tricked and robbed of their just reward. Crompton received a mere £60 for his spinning 'mule', but Boulton made quite sure that his partner was not similarly swindled.

By the end of 1776 a third engine had been built, this time in London for a distillery at Stratford-Le-Bow. James Watt's own drawing of this engine still survives, and it shows very clearly the layout and design of this relatively small beam engine with its 18-inch-diameter cylinder. A feature of this cylinder was a steam-jacket to keep it hot. Steam was circulated between double walls – an idea previously tried out by Watt on his model.

During the next fifteen years Watt, who was not content to rest on his laurels, produced a whole series of improvements in the design of his steam engines. In these early days practically every engine included a new idea. Some of these were successful and retained in later designs, but others were discarded.

A large engine for a colliery near Coventry had a new type of valve to control the flow of steam into and out of the cylinder. The earlier engines had a single sector valve, similar to the one used by Newcomen, but Watt introduced separate 'drop' valves which operated in the same way as an ordinary plug in a wash basin. This innovation was one of those retained.

The year 1776 was a busy one for Watt with his first engines erected and working, as well as designs for new ones. Even so, in June he journeyed north to Scotland to arrange his wedding. Watt's first wife had died several years earlier and he was now forty years of age with a considerable reputation as an engineer. His prospective father-in-law was not impressed and demanded to see Watt's contract with Boulton. This was an embarrassing request and Watt hastily wrote to his partner:—

> '... I find that the old gentleman wishes to see the contract of Partnership between you and I, and as that has never been formally executed, I must beg the favour of you to get a legal contract written and signed by yourself, sent by return of post or as soon as may be ...'

Watt and Boulton had such complete trust in each other that they had not resorted to a legal partnership, but the enterprising Boulton rescued his partner with a letter explaining that the document could not be seen because their lawyer was in London, but enclosing a fictitious summary. The white lie satisfied the father and the wedding went ahead.

The personal nature of his work in Scotland did not deter Watt from doing business, and he obtained the first order for one of his engines in Scotland. It was to be built at Torryburn, near Dunfermline. A feature of this engine was the furnace under the cylinder – yet another attempt to keep the cylinder hot. Experience showed that this was no improvement on the steam-jacket already tried, and Watt discarded the idea.

The news that Boulton and Watt engines were using one third of the coal burnt by a corresponding Newcomen-type engine led to many orders – especially from the Cornish tin mines. Coal was expensive in Cornwall and the mines were deep and wet, making pumps vitally important. Incidentally, Boulton and Watt did not sell their engines outright: the owner paid for the parts and the cost of erection, but he also paid a royalty equal to one-third of the money he saved by using a Boulton and Watt engine instead of a Newcomen-type of similar power. Because their engines burnt so much less coal, the royalty brought Boulton and Watt a steady income which was to continue until the patent expired in 1800.

In 1777 Watt's fertile mind was busy working on two new engines, not just modifications but new concepts. Very little is known about the first, which was supplied to the ironworks of John Wilkinson – who had one of the first pair of Boulton and Watt engines. The new engine was not a beam engine; instead its cylinder was turned upside-down, thus making it possible for the piston rod to be extended downwards and drive a water pump immediately below. Because of its inverted layout it became known as the 'Topsey Turvey Engine', but no drawings or details of its performance survived. We can only assume that it was not a great success because, despite its compactness, the design was not repeated.

The other new type of engine was powered by steam at a higher pressure than usual, and its piston was moved by the action of the steam expanding. As mentioned earlier, Watt was not enthusiastic about high pressures; nevertheless in 1777 he tackled the problem, perhaps spurred on by Boulton who was enthusiastic about the idea. An engine was erected at the Soho works of Boulton and Watt for their own use, so that trials and experiments could be carried out privately. The 'expansive' engine was not a great success, but on the other hand it was not a failure. Because Watt preferred his conventional low pressure engines, perhaps he did not give the new design enough attention. Certainly the workmen found it rather a handful to control and it was nicknamed *Beelzebub*. As the years went by *Beelzebub*'s violent ways were brought under control and its nickname changed to the more affectionate *Old Bess*. In 1848 *Old Bess* was retired from the Soho works, and the major parts of this famous engine are now displayed at the Science Museum, South Kensington. Another Boulton and Watt engine built in 1777 is preserved in the Birmingham Museum of Science and Industry.

James Watt was very busy designing and supervising the installation of beam engines to pump water, particularly for the tin and other mines in Cornwall. Matthew Boulton, on the other hand, was looking to the future, for by this time the Industrial Revolution was in full swing and factories needed power. He tried to persuade Watt to produce an engine which could rotate a shaft in order to drive machines. Watt experimented with models, but in the meantime a rival company took out a patent which included the crank. This was a strange business, and the truth is difficult to determine as both sides tell a different story, but it does seem odd that a patent was allowed for a crank, which was a well-known method of driving a wheel. On the other hand, no one had fitted a crank to a steam engine.

In the year 1779 the rival company of Matthew Wasbrough and James Pickard built a Newcomen-type engine at Snow Hill, Birmingham, and they fitted it with a pawl and ratchet device to produce rotary motion. It aroused much interest, but its movement was far from smooth and trouble-free. The following year a mechanism was fitted which included a crank and the idea was patented. One of Watt's workmen admitted that he had mentioned his master's experiments with a crank to Pickard's engineman while they were having a drink together in Birmingham. But Pickard claimed it was his own idea, while Watt said it had been stolen from him.

Boulton and Watt did not contest the patent; instead Watt – perhaps with the assistance of his employee William Murdock – solved the problem by devising a system of gear wheels which Watt patented late in 1781. This gearing was called the 'sun-and-planet motion', because one gear, the 'planet', actually moved around a central 'sun' gear. The 'planet' gear was attached to the lower end of the connecting rod and it was guided around its 'sun' by a 'free-wheeling' crank. Unlike Pickard's crank, Watt's did not drive the shaft directly; instead the crank was free and the drive was transferred through the 'sun' gear. This arrangement not only avoided Pickard's patent, but also had another advantage, in that it increased the speed of the drive shaft and flywheel. One of the beam engine's inherent faults when used to power machinery was its very slow speed. A piston would make only about twenty strokes per minute, which the sun-and-planet gear could convert to perhaps forty revolutions per minute at the driving shaft.

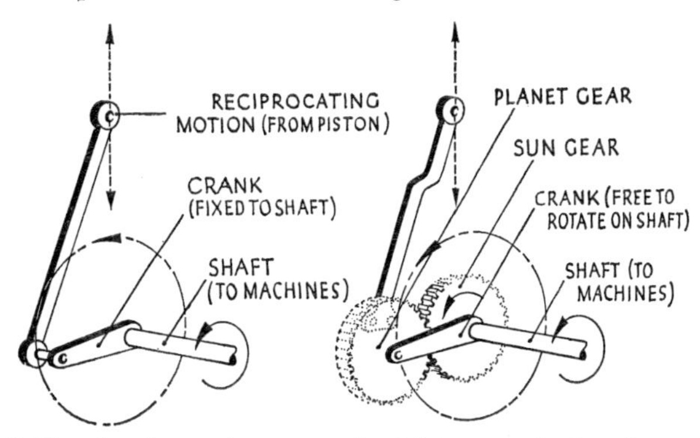

FIG 18 *The simple crank compared with 'sun-and-planet' gearing.*

Following an earlier sequence of events, Boulton and Watt built their first rotative beam engine at Soho for their own use, and the second one went to their associate John Wilkinson. The Soho engine first ran towards the end of 1782, and Wilkinson's engine, which powered a mechanical hammer, followed in March 1783.

The rotative engine was still far from perfect. In the previous chapter the jerky movement of the Newcomen-type engine was explained: 'It was single-acting, which meant a power stroke in one direction only, followed by a return stroke due to gravity acting on a counterbalance weight.' While Watt had improved engine power and efficiency, the basic mechanical layout remained the same. A heavy flywheel helped to smooth out the jerks, but Watt had other ideas. He decided to make both the up and the down strokes of the piston into working strokes, thus giving a more even distribution of power. Instead of the piston driving the engine for less than half a complete cycle, it was driving almost all the time. Watt patented his double-acting engine in 1782 and had a trial engine running at Soho in the following year.

A new problem was created by the double-acting engine, because a simple chain could no longer be used to connect the piston rod to the beam. In a single-acting engine the piston was powered only on the downward stroke and a chain was satisfactory, but with a double-acting engine the piston also pushed upwards which meant that a rigid connection was required. As already mentioned, a direct connection was not possible because the end of the see-saw beam moved in the arc of a circle, while the piston rod had to move in a straight line. Watt's first solution was to fit part of a giant gear wheel at the end of the beam and engage its teeth with those of a rack (or bar with teeth) which was attached to the vertical piston rod. This rack and sector was tried out on the Soho engine, but it was noisy and frequently broke down.

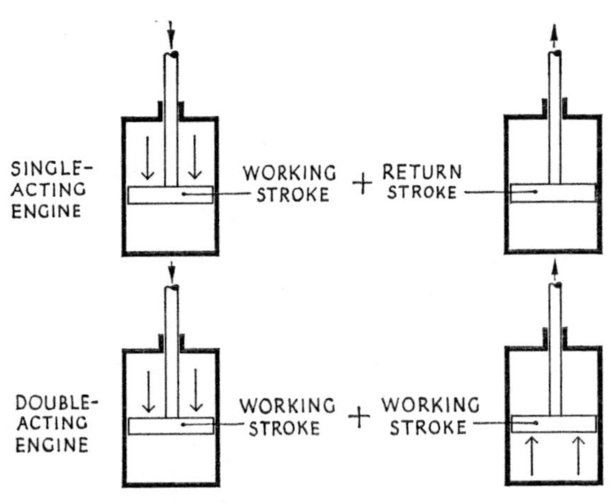

FIG 19 *Single- and double-acting engines compared.*

The first double-acting rotative engine for a customer was designed in June 1784 with a rack and sector, but before the engine was erected in Hull this mechanism had been discarded in favour of a system of links. In principle this used three links arranged in the shape of Z (three pieces of Meccano would demonstrate this more clearly). The piston rod was attached to the mid-point of the upright of the Z. As this point moved up and down it did so in a vertical straight line – just as required. In practice the upper horizontal link was one side of the beam of the beam engine. Several engines were built with this three-link motion, but Watt was still not satisfied because the links were long and cumbersome, as they had to be in order to cope with the movement of the piston.

Watt's next patent was the one of which he, personally, was most proud. It overcame the problem of long links by reducing their size to a convenient length and then transferring the straight-line motion to the piston rod by means of a system of parallel links, which magnified the length of the straight line to suit the stroke of the piston. The same system of links can be used as a 'pantograph', which produces an enlarged copy of any drawing by following the original shape with a pointer. Watt patented his parallel motion in 1784, and the first engine designed to incorporate it was, rather surprisingly, one which was only single-acting. This engine was built for Whitbread's brewery in London and is now preserved in Sydney, Australia.

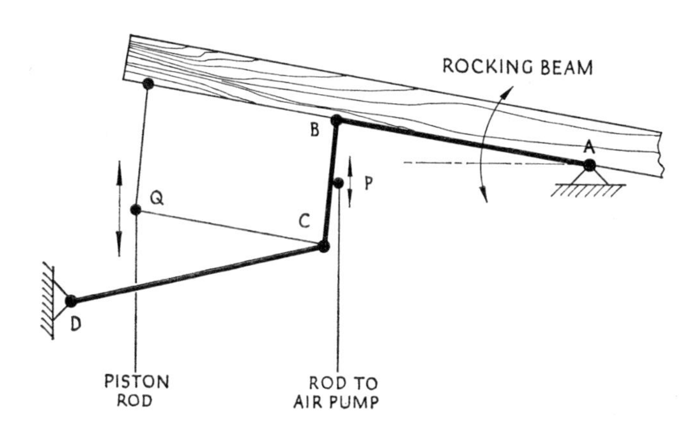

FIG 20 *Watt's parallel motion. The three links* ABCD *guide point* P *to move in a vertical straight line. The remaining links form a pantograph, which magnifies the movement of* P *at point* Q*—to which the piston rod is attached.*

Watt's parallel motion became very widely used for all types of beam engines and remained popular until machine tools could produce the long true flat surfaces required by the alternative – a sliding crosshead to guide the piston rod.

While beam engines were being used to pump water, it was not essential that their speed should be kept steady, but when machines in factories were being powered it became obvious that a constant speed was required. In 1787 Watt tackled and solved this problem by adapting an existing device called a conical pendulum governor. The principle had already been invented to control the speed of windmills and waterwheels, and therefore Watt did not take out a patent although it is often called a Watt governor. It consisted of two hanging rods, each with a spherical weight attached at the lower end, but whereas the weight of a clock's pendulum swings to and fro, these weights moved in a circular path around a vertical spindle. They were driven round by a belt or chain from the main driving shaft of the beam engine. If the speed increased, then the weights moved outwards due to the extra centrifugal force, and this movement was cleverly transmitted by links to the engine's steam supply valve, which was closed a little. In this way, if the engine ran too quickly, the steam supply was reduced, and if the speed dropped below the limit, the valve opened to admit more steam.

The first engine to be fitted with a governor was a double-acting beam engine erected at the Soho works, where it drove machinery for lapping or polishing the ornaments for which Boulton was famous. It became known as the 'Lap' engine and it also can be seen in the Science Museum, South Kensington. This engine represents the virtual end of Watt's period of steam engine improvement; he had increased the 'duty' or efficiency from under 10, the value achieved by Smeaton, to approximately 30 (i.e. 30 million pounds of water raised one foot per bushel of coal burnt). Boulton and Watt concentrated on producing engines based on Watt's basic design, for their patent protected them from competitors until 1800. This protection was not completely effective, and the partners had several legal battles with rivals to protect their rights.

The concept of the beam engine did not change very greatly in fifty years. There were relatively minor changes, which included different valves to control the steam, cast-iron beams instead of wooden ones, higher pressures and cranks in place of the sun-and-planet gears. The separate condenser, however, with its air pump, parallel motion, double-acting piston and governor remained.

Although Boulton and Watt were selling new engines throughout the country, they also rebuilt a number of their early engines to include the later inventions. For example, Barclay & Perkins' brewery in London installed a Boulton and Watt engine during 1786 for pumping water and grinding malt. This was a typical engine of that period, single-acting (double-acting was still new and suspect), rotative with sun-and-planet gearing and parallel motion. In 1796 it was rebuilt and converted to double-acting, with a larger condenser and including the latest modification – a governor. This engine is preserved at the Royal Scottish Museum, Edinburgh.

Calculating the size of an engine to pump water was relatively simple, but when waterwheels and horses were being replaced the problem was more complex. For example, the engine for Barclay and Perkins was designed not only to pump water but also to replace a horse gin – a 'roundabout' to which horses were harnessed. Two horses walked round and round, thereby turning several mill-stones which were connected to the gin by means of a shaft and gears. Other engineers, including Smeaton, had measured the power of engines by the number of horses they replaced, and a few even calculated how much work a horse could do. If the engine was to replace several horse gins the only problem was to calculate the size of engine to produce this power, but if a new factory was to be powered, the power required had first to be calculated.

Watt studied this question of horse-power, but he did not make any pronouncement; instead it just filtered into his calculations. In 1782 his calculations, relating to a beam engine for a Mr Worthington of Manchester, show how he used the fact that twelve horses were being replaced to calculate the size of the engine's cylinder. This calculation indicates that one horse was assumed to be capable of working at the rate of 32,400 pounds raised one foot per minute. (Note: 'Work done' is load × distance lifted; thus 32,400 lb. × 1 foot could have been 32·4lb. × 1000 ft, and 'power' is the rate of doing work.) The following year Watt's calculations show he was using a figure of 33,000 pounds – feet per minute for one horse-power, and this figure is still used today. By a strange twist of history the unit of power which carries Watt's name is not the one he virtually defined, but an electrical unit introduced long after his death.

The Soho Manufactory achieved world-wide fame, not only for steam engines, but also because of the high-quality goods Boulton produced, which varied from ornate silver candelabra to buckles, coins and even buttons. Naturally, a company of this standing attracted some fine workmen, and perhaps the most famous of these was a Scottish engineer and inventor named William Murdock. Incidentally, his parents spelt their name Murdoch, but in England William was generally referred to as Murdock: certainly Boulton and Watt spelt his name with a 'k'.

Murdock is often labelled 'the inventor of gas lighting' but, as so often happens, the word 'inventor' is misleading. He was a great pioneer of coal-gas lighting, and if the process of invention is divided into five stages (as mentioned in an earlier chapter) Murdock contributed to the fourth stage with a practical demonstration, first at his house, then at the Soho Works. But he was not the first man to demonstrate gas lighting and he did not take it into the final stage by selling it at an economic price. One of the first gas engineers who produced this new source of power on a commercial scale was Samuel Clegg – yet another former Boulton and Watt employee. At this time gas was being developed primarily for lighting purposes, but in later years it became a fuel for the early internal-combustion engines.

William Murdock joined Boulton and Watt in 1777 when he was twenty-three years of age and for a short time he worked at the Soho Works. Being a very intelligent young man, he soon acquired a detailed knowledge of Watt's engines and in 1779 he was entrusted with the erection of an engine at Wanlockhead, in Scotland. No sooner had he returned to Birmingham than he was sent to Cornwall to supervise the erection of more Boulton and Watt engines. Here Murdock stayed for almost twenty years and established a great reputation as an honest, hard-working, reliable man and a brilliant engineer.

In 1781 Murdock assisted Watt with the 'sun-and-planet motion' and frequently he indulged in inventions of his own. While some of these involved improvements to beam engines, a number were not connected with the interests of his employers. Watt did not approve of his own designs being altered by Murdock, and both partners were apprehensive about any original inventions, just in case they were successful – for they did not wish to lose Murdock's services.

Murdock produced two important inventions using steam-powered models during the mid 1780s; the actual dates are not known and neither was developed by the company. One consisted of a rotative steam engine which eliminated the cumbersome overhead beam, and the second was a model of a steam-powered vehicle which is described in a later chapter. The former was called an 'oscillating

PLATE 10 *A typical horizontal steam engine of the mid-nineteenth century. Until recently this engine powered railway workshops. This illustration gives some idea of the difficulty of photographing steam engines in dark, confined engine houses.*

PLATE 11 *A copy of Murdock's model vehicle with a miniature steam engine.*

PLATE 12 *Henry Bell's famous steamboat, the* Comet *of 1812.*

PLATE 13 *The* Wylam Dilly, *built by William Hedley in 1813.*

PLATE 14 *A model of Stephenson's famous* Rocket *of 1829.*

PLATE 15 (*Top*) *A Crampton locomotive; the* Kinnaird *of 1848 was built for the Dundee and Perth Railway.*

PLATE 16 *Sir Hiram Maxim's steam-engined aeroplane of the 1890's.*

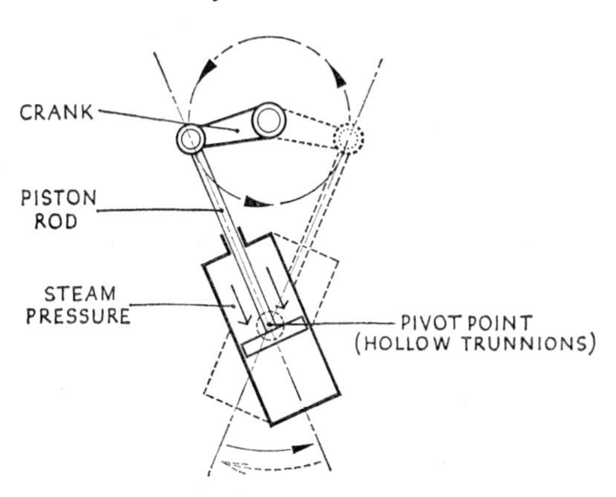

FIG 21 *A simplified oscillating engine showing how the cylinder rocks to allow the piston rod to be connected directly to the crank. Steam enters the cylinder through the hollow trunnions and then the valves.*

engine' because it had a cylinder which rocked, thereby allowing its piston rod to be connected directly to a crank. This was a very simple layout which did not require a beam or parallel motion linkage. However, the difficulty of piping steam to a cylinder which rocked had still to be overcome. The rocking cylinder was supported on trunnions, rather like an old-fashioned cannon, but it was 'aimed' upwards because the crank was above the cylinder. Murdock's solution to the problem of supplying steam to the moving cylinder was to feed it through one of these trunnions which was made hollow. The model worked, using compressed air in place of steam, but it was left to later inventors to carry this idea into a full-size engine.

Murdock experimented with a number of other ideas while in Cornwall, including gas lighting, as already mentioned, and the use of compressed air to transmit power, but his greatest contribution to the steam engine was a new steam valve and operating mechanism. Even on his model steam carriage Murdock had not used Watt's plug-type valves; instead he had used miniature pistons which moved up and down inside a tube, thereby revealing and covering holes into the cylinder. Murdock's new valve, developed from this idea, was not, however, used until the end of the century.

By 1794, both Matthew Boulton and James Watt were nearing the age of retirement and their sons had virtually taken over. Very wisely the young men decided to make use of Murdock's ideas and experience; thus in 1798, at the age of forty-four, he was transferred from Cornwall to the Soho Works where he eventually became a partner in the Company. Soon after moving to Birmingham, Murdock patented several of his inventions, one of which was a new type of valve to control the flow of steam to a cylinder. It worked rather like a sliding door, but it was not a flat 'door', being shaped something like a letter C – or, if viewed from the reverse side, a letter D. Although the former was perhaps a better description, it became known as the D-slide valve.

The D-slide valve moved to and fro allowing steam to flow first to one side of the piston then to the other. Its action can perhaps be understood by considering three holes in a row and an open box placed, with its open end down, over two of the holes. Steam comes from the middle hole into the box and out through one end hole. Now, if the box is slid along until it is over the middle hole and the other end one, steam is cut off from the first end hole and allowed into the second. In this way steam is supplied to each end of a double-acting cylinder in turn.

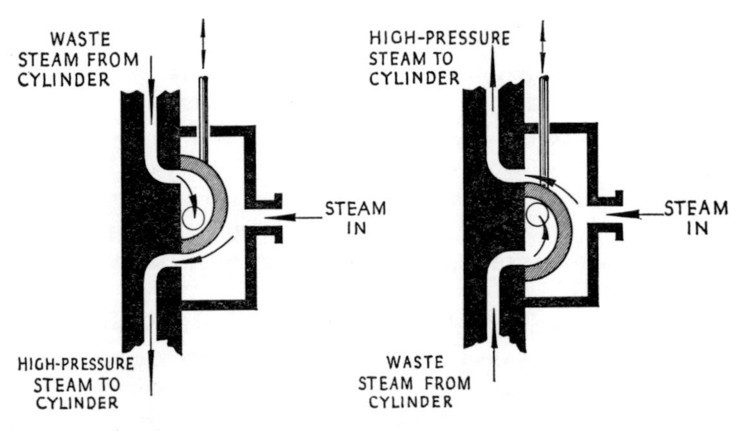

FIG 22 *The D-slide valve used to control steam flow to a cylinder. High-pressure steam is directed first to one side of the piston, then the other.*

The reciprocating movement of a D-slide valve could be obtained by connecting it to the parallel motion links of a beam engine or by including an extra crank, but the ingenious Murdock invented another device which enabled his valve to be used for engines without beams. An extra crank on the crankshaft would have been bulky, heavy and expensive in view of the small power and movement required to operate a valve. Murdock introduced a mechanism called an 'eccentric'. A circular metal disc with a hole in it was slid on to the crankshaft and locked so that it rotated with the shaft. A loose-fitting ring was placed around this disc and connected, by a rod, to the slide valve. The secret was to make the hole in the disc off-centre, because this offset had the same effect as a crank, resulting in the valve rod being moved to and fro. A more technical explanation would be to compare the eccentric with a cam. The cam follower (valve rod), is held in contact with the cam (disc) by means of a strap around the cam (loose-fitting ring). Of course with most cams the follower is held in contact by a spring. The D-slide valve operated by an eccentric became standard practice for steam engines and remained in use for a hundred years or more.

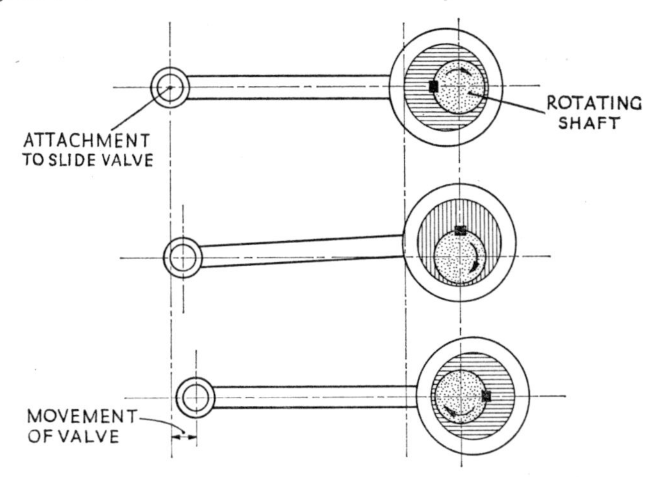

FIG 23 *An 'eccentric' mechanism used to operate a D-slide valve. As the shaft rotates the attachment point to the slide valve is moved, first to the right (as shown) then back to the left.*

During the last decade of the century Boulton and Watt offered their customers a series of standard engines ranging from 10 horse-power to 50. A 10 h.p. engine had a cylinder about 18 inches in diameter, while the 50 h.p. engine was just double this diameter. Business was good and by the year 1800 the number of engines sold had risen to approximately 500. In the same year the patent expired and Watt, then aged sixty-four, retired in favour of his son.

CHAPTER 5

High-Pressure Steam Engines

1800–1850

When James Watt retired in the year 1800 it was the end of an era. The old master had dominated the scene for a quarter of a century, and it was up to his followers, and rivals, to move ahead. During the later years Watt had been so engrossed with his successful beam engines that eventually he became a brake on progress. His own workers were not encouraged to waste their time experimenting, and rivals were threatened with the patent. Some of these men were very able engineers, and so the first quarter of the nineteenth century was a period of exciting experiments both in the design of the steam engine and in its application.

Probably the greatest contribution to the steam engine story during this period was made by a Cornish engineer called Richard Trevithick, who worked for one of Watt's rivals, Edward Bull. Despite this, Trevithick had tried to join Boulton and Watt, but his application was turned down, probably on the advice of Murdock who knew Trevithick in Cornwall. Richard Trevithick became one of the greatest inventors, but much of his talent was wasted because he did not have a friend like Matthew Boulton.

Trevithick was an advocate of high-pressure steam to drive his engines, and their usual limit was 50 pounds per square inch. Watt, it will be remembered, used low-pressure steam at less than 10 pounds per square inch and obtained power by condensing the steam. In Trevithick's engine the piston was moved by the pressure of the steam, and at the end of each stroke the steam was exhausted into the air without condensing. Trevithick was restrained by Watt's patent until 1800, but in 1802 he was able to patent his own high-pressure engine, and in the same year he built an engine which worked at the phenomenal pressure of 145 pounds per square inch.

Trevithick made another great step forward by discarding the traditional see-saw beam in favour of a 'direct acting' layout for many of his engines. The free end of his piston rod was attached to a crosshead which was constrained to move in a straight line by guide rails. A connecting rod joined this crosshead to the crank, making a direct system of links from piston to crank. One of Trevithick's high-pressure engines and boiler is displayed in the Science Museum, South Kensington. These engines were very compact compared with the older beam engines and they proved to be successful as winding engines for mine cages. It was this compactness which gave Trevithick a chance to power a vehicle, while the sheer bulk of Watt's beam engine made it unsuitable for this purpose.

Although the condensing beam engine was large, it could be used as a stationary engine to power machinery, and in this rôle its popularity grew. Even as early as 1781 Matthew Boulton wrote to James Watt saying, 'The people in London, Manchester and Birmingham are steam mill mad'. Perhaps he was exaggerating a little to encourage Watt, who at that time was experimenting with the rotative engine which later superseded the waterwheel. By the end of the century steam engines had been installed in mines, ironworks, flour mills, oil mills, breweries, a starch manufactory and several other mills and factories. After 1800 even more industries turned to steam for power – paper-making and printing, for instance. In 1814 the first newspaper to be printed by a steam-driven press was published. The cotton mills of Lancashire and the woollen mills of Yorkshire were providing a good market for steam engines, and many new manufacturers set up in business in these and other industrial areas.

The spread of steam power into the field of transport is described in a later chapter, but perhaps just as important as the spread of steam to new industries was the overall increase in the number of engines being used. As engines became more efficient their popularity grew, and the quest for higher efficiencies produced many new designs. Some of these were soon forgotten, but others remained for a hundred years or more.

The general trend towards greater efficiency continued throughout the nineteenth century. During the first half, the beam engine dominated the scene, challenged occasionally by the direct-acting engine, but these positions were reversed in the second half of the century. To simplify the description of these alternative layouts, their stories for the period 1800 to 1850 are told separately.

BEAM ENGINES

Prior to the year 1800 customers could choose only a Newcomen or a Watt type of beam engine, but after the turn of the century the selection grew considerably. Yet despite the almost universal demand for higher efficiency one group continued to buy the outdated Newcomen-type of engine. They were the colliery owners of the north-east of England who preferred the old engine because it was simple in operation and inexpensive to buy. The fact that it was inefficient and burnt large quantities of coal did not worry them because, as colliery owners, they had plenty of waste, second-rate coal to burn. By contrast, owners of tin mines in Cornwall had to pay a high price for coal, because it was not produced locally and transport was very expensive.

The most noticeable change in beam engine design during the early years of the nineteenth century was probably the change-over from wood to iron. A typical Boulton and Watt engine had a great wooden beam, supported by a massive timber framework which included supports for the various parts of the engine. An engine house some 30 feet high was necessary, with the result that a building had to be built around the engine. The size of engines for any particular horsepower was gradually reduced, partly by the designing of more efficient engines and partly by the use of cast iron. Because iron is considerably stronger than wood, a smaller beam could be used, which in turn led to a reduction in the supporting framework. In some cases the supporting framework was reduced to a single cast-iron column under the pivot point of the beam. These columns were often decorated with 'flutes' along their length and with ornate tops and bases. A further improvement was the introduction of a cast-iron 'bed plate', or base, for the engine and all its components. Although this did not make the engine portable, it could, at least, be erected in any convenient building without major structural alterations.

One of the first great engine-builders of this post-Watt era was Matthew Murray of Leeds. Not only were his designs sound, but the standard of workmanship in his workshops was extremely high. Murray simplified and lightened Murdock's D-slide valve in 1801 and a few years later he brought out a 6 horse-power portable beam engine. Unfortunately it was not a great success and only a few were built.

Although Trevithick spent much of his time designing direct-acting engines he did not neglect the beam engine. His first high-pressure engine was a beam winding engine for a Cornish mine, and in 1802 he astonished rival engineers by building a beam engine which worked at a pressure of 145 pounds per square inch. Even Trevithick's usual figure of 50 was thought to be rather high by the more conservative designers. This extra high-pressure engine was used for pumping water at Coalbrookdale, and its small size compared with other engines of similar power was very noticeable. Though the engine was small, the boiler had to be more substantial than usual to withstand the pressure. It was made in cast-iron with walls 1½ inches thick. By 1804 nearly fifty of Trevithick's engines had been constructed, including both beam and direct-acting designs using a variety of layouts.

Trevithick had a counterpart in the United States who advocated the use of high-pressure steam; he was Oliver Evans, a flour-miller by profession. Evans almost re-invented the steam engine by himself, because even as late as the year 1800 there were only half a dozen steam engines in the United States, and it is doubtful whether Evans saw one of these. He started thinking about the power of steam after watching some young friends heating water in a gun barrel which had a plug in the open end. The plug was blown out with an impressive bang, much to the amusement of his friends, but Evans began to think in terms of a cylinder instead of a gun barrel and a piston instead of the plug. Before he could design a high-pressure engine, Evans had to develop a suitable boiler, but he persevered and by 1805 he was planning to use a pressure of 120 pounds per square inch.

About the same time he designed a half-beam engine which became known as a 'grasshopper'. It is interesting to note that this type of engine was invented almost simultaneously in England by William Freemantle, without any known contact between the two inventors. The 'grasshopper' beam was more compact than the normal 'see-saw' because it was effectively only half a beam, with the cylinder and crank positioned on the same side of the pivot. It had a second, and in some respects a more important, advantage in that it did not require the complicated parallel-motion linkage; nor did it use the relatively expensive sliding-crosshead to guide the piston rod in a straight line. The key to this clever arrangement was the support for the pivoted end of the half-beam, which was a vertical swinging link and not a rigid support. This relatively simple mechanism guided the piston rod along an almost straight line, and

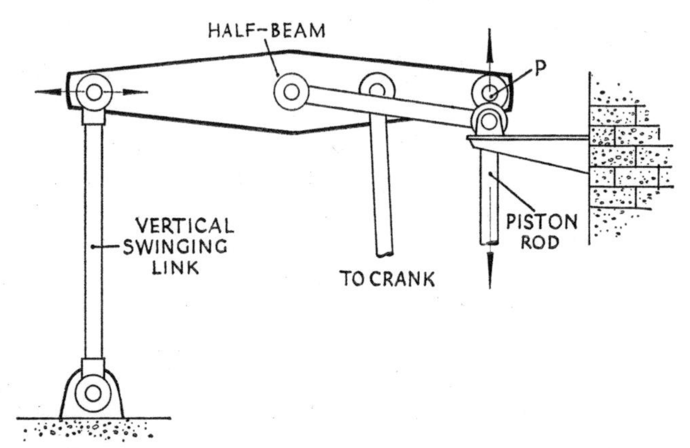

FIG 24 *The 'grasshopper' principle, showing the components which guide the upper end of the piston rod* P *along a vertical straight line.*

the beam moved with a peculiar loping action – whence the name 'grasshopper'.

William Hedley's locomotives *Puffing Billy* and *Wylam Dilly* of 1813 used this variation of the beam engine. In fact they were even more like grasshoppers because they each had two loping beams, one for each cylinder on either side of the boiler.

The 'grasshopper' proved to be a very useful design, particularly for smaller engines. They were used to pump water, power machinery and accomplish many other tasks. One was installed at the Royal Albert Hall in London to pump water up from a well below the building. But as a design they did not represent a step forward in the way they utilized steam. Perhaps the greatest advance in this field was the 'compound' engine, a two-stage arrangement in which steam first passed through a high-pressure cylinder, then a low-pressure cylinder before it was exhausted. Naturally this principle required a high initial pressure, and it was the failure to achieve this which resulted in the early compound engines having only a limited success.

The idea of compounding an engine was patented as early as 1781 by one of James Watt's rivals, Jonathan Hornblower – a member of the famous family which has already been mentioned. Jonathan's father, also called Jonathan, erected Boulton and Watt engines in Cornwall, and his grandfather, Joseph, was an associate of Newcomen. Young Jonathan's first compound engine was erected at Radstock Colliery, near Bath, and by the end of 1782 it was working

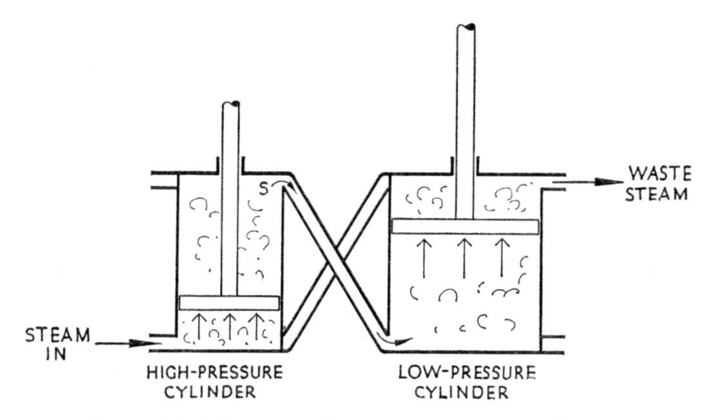

FIG 25 *A simplified diagram of a compound engine.* S *represents low-pressure steam from the previous working stroke being transferred to the larger low-pressure cylinder for a second working stroke. This movement is controlled by valves (not shown).*

satisfactorily. Its cylinders, which were of 19 and 24 inches diameter, were both attached to the same beam by chains; this was possible because they were only single-acting. At about the same time Watt doubled the power-output from a single cylinder by making it double-acting (i.e. a working stroke in each direction) and this virtually eliminated the advantage of Hornblower's design. Despite this he managed to sell about nine or ten in Cornwall.

To be effective, the compound engine had to have high-pressure steam, which required a reliable boiler. In 1803 a patent for a new type of boiler was taken out by Arthur Woolf, and the following year he added a high-pressure cylinder to an existing Watt engine at Meux's Brewery in London. Unfortunately, Woolf made the new cylinder too small and there was no improvement. Woolf left the brewery, but he did not lose faith in the idea and entered into partnership with another engineer called Humphrey Edwards, who had a workshop in Lambeth. By 1811 they were building compound beam engines which were really effective and consumed only half as much fuel as a simple Watt engine of the same power. The layout of Woolf and Edwards' engine was the traditional beam engine pattern, but with a high-pressure cylinder placed alongside the normal low-pressure cylinder. Both piston-rods were connected to the same pin of the parallel motion.

In 1811 the partnership of Woolf and Edwards broke up. Edwards later moved to France where he had previously sold about a hundred compound engines. He joined a French Company which built several hundred more, following Woolf's general design. Woolf himself moved to Cornwall where he erected a number of large compound engines working at pressures of up to 40 pounds per square inch. These were reasonably successful at first, but as the efficiency of simple steam engines increased, the advantage gained by compounding was lost. The compound engine in Britain disappeared – to be re-invented some thirty years later.

One of the reasons for the failure of Woolf's compound engine to sell in large numbers in Cornwall was the introduction of a new variant of the simple beam engine there (when describing steam engines the term 'simple' is often used to denote non-compound). In 1812, Richard Trevithick returned to Cornwall, having failed to make a financial success of his experimental steam carriages and locomotives. He turned his attention to the beam engines used for pumping water out of the mines. Many of these still worked on Watt's principle, which gained most of its power from the vacuum created by condensing steam, although very low-pressure steam was used to assist in moving the piston. In other words, the piston was moved by 'sucking' on one side and 'blowing' on the other, the 'blowing' being very light. Trevithick decided to increase the pressure but retain the suction.

Using high-pressure steam to move a piston in a cylinder sets a problem for the designer. Maximum power is obtained by allowing steam to enter the cylinder during the whole time it takes the piston to move from one end to the other (i.e. the stroke); but maximum efficiency is achieved by allowing steam to enter the cylinder for only a fraction of the stroke. When the steam inlet valve is closed, the trapped steam still exerts a pressure on the piston, which continues to move. With the movement of the piston, this pressure drops because the steam is expanding to fill a larger space. Now, the problem for the designer is to achieve a balance between high power with low efficiency and low power with high efficiency.

James Watt had used expanding steam in 1777 with his engine *Beelzebub*, but he never really favoured high pressures. When Trevithick designed his high-pressure beam engine in 1812 he had more than ten years' experience with high-pressure boilers and engines. He used a cylindrical boiler in which the fire burned inside a large tube surrounded by water – a type of boiler which became very popular and known as the 'Cornish' boiler. The engine, too, became known

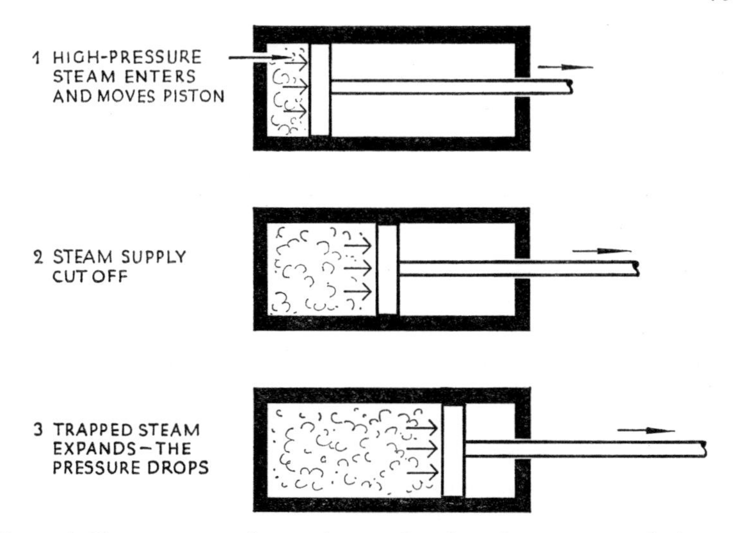

FIG 26 *How an expansive engine works when the steam supply is cut off before the end of the stroke.*

as a 'Cornish' pumping engine. Trevithick's engine was supplied with steam at 40 pounds per square inch, which could be cut off after only one tenth of the stroke. By adjusting the cut-off point, more, or less, power could easily be obtained. Trevithick did not develop this high-pressure plus condensing engine, but others did with great success.

In an earlier chapter the 'duty' of engines was described; this was a measure of efficiency in terms of millions of pounds of water raised one foot while consuming a bushel of coal. A good Watt engine could almost achieve 40 and Woolf's compound engine reached a figure of 68. By the 1830s Cornish engines were attaining a duty of 100, and some operators claimed 125 – though 'rigged' trials were not unknown. This efficiency, combined with a remarkable reliability, resulted in Cornish engines being used for over a hundred years. In fact, one was re-erected for use at a Cornish mine as late as 1924. Because many of these magnificent engines remained in service until recent years it has been possible to preserve several, but because of their size they are better preserved 'on site' – a task which has been undertaken by the Cornish Engines Preservation Society.

The typical Cornish engine was inherently large for several reasons. An early cut-off point for the steam resulted in high efficiency but low power; therefore to increase the power meant increasing the size of the engine. Cornish engines were normally single-acting, which, again, resulted in a large cylinder compared with one which was double-acting. Cylinders 7 feet in diameter, with a stroke of up to 12 feet, were quite common, and these proportions naturally resulted in an enormous beam. Cast iron was used to manufacture these beams, which weighed as much as 30 tons. The working stroke was normally the down stroke of the piston and this coincided with the up, and pumping, stroke of the pump attached to the other end of the beam. The piston was returned to the top of the cylinder ready for the next working stroke by making the pump end of the beam heavier – just as Newcomen had done a century earlier. This was easily arranged in the mines because the pump was far below ground level, so that the weight of the rods extending down to the pump provided the necessary weight. In some cases, when the mine was very deep, the weight of these rods was excessively high and had to be counterbalanced by a separate beam and counterbalance weight.

Although these rather slow and ponderous Cornish engines had their limitations, they were ideal for pumping large quantities of water. As well as being introduced in to the mines of Cornwall they were installed at many water works in different parts of the country during the mid-nineteenth century.

Beam engines were bulky, and in particular high, which was a considerable handicap when they were installed in ships. Nevertheless, many conventional beam engines were built into American paddle steamers during the 1820s and some remained in service for seventy to eighty years. British marine engineers favoured a modified beam engine known as a 'side-lever' engine. Instead of one beam this had two, positioned low down near the base of its vertical cylinder, one on either side. The piston rod still emerged from the top of the cylinder, but it terminated in a crosshead from which two links ran back down the sides of the cylinder to the twin beams. These beams were pivoted in the centre, and from their other end a connecting rod turned a crank and hence the paddles.

The side-lever engine was very heavy, but the weight was in the right place for a ship – low down below the waterline, thus improving the vessel's stability. It was installed in many of the early Clyde-built steamers during the first half of the nineteenth century and even as late as the 1860s. But the side-lever never had much success as a source of power outside marine engineering.

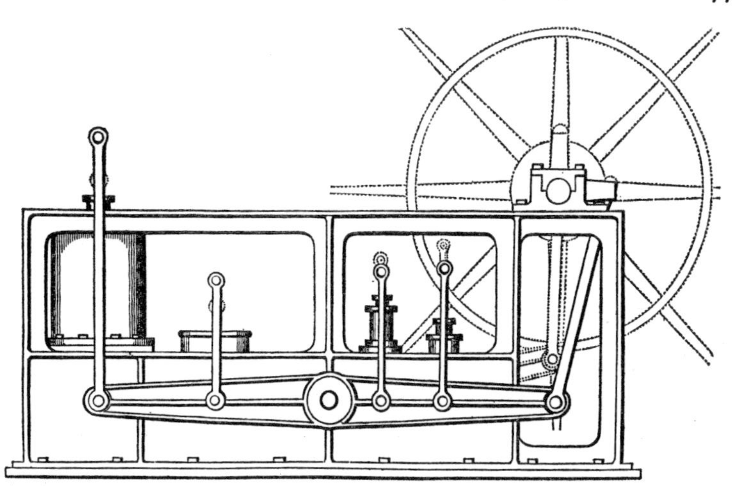

FIG 27 *A side-lever engine with two levers or beams, one on either side of the framework containing the cylinder, air pump etc. One of the paddle-wheels is shown by dotted lines, on the right-hand side of the illustration.*

Even as late as 1845 there was a revival of interest in the beam engine, when a really successful compound engine was produced by an engineer called William McNaught. Not only did he build efficient compound engines; he also devised a most ingenious layout which made it possible to convert existing beam engines into compound engines. So successful was this arrangement that the conversion was called 'McNaughting' the engine. A high-pressure cylinder was added, but instead of placing it adjacent to the main cylinder McNaught connected it to the beam mid-way between the pivot and the connecting rod (i.e. on the other side of the 'see-saw'). The great advantage of this position was that the extra power generated by the new cylinder did not have to be carried across the pivot point – the most highly loaded part of the beam. In this way more power could be obtained from existing engines without overloading the beam. Some new engines were also manufactured to McNaught's design, and these were supplied with steam at a pressure of 120 to 150 pounds per square inch.

It is clear that, by adding a cylinder half-way between the pivot and the end of the beam, the stroke of this piston will be less than that of the piston attached to the end of the beam (a half, in fact). But this is not a bad feature, because high-pressure steam occupies much less space than low-pressure steam, and therefore a high-pressure cylinder must be smaller in diameter or stroke or both.

Compound beam engines were also designed with both cylinders on the same side of the beam and these were known as Woolf compound engines because they used Arthur Woolf's layout of 1804. Once again the smaller, high-pressure cylinder was nearer the pivot point. Naturally these beams had to be designed to carry the combined power across the pivot, to the connecting rod and crank, but the close proximity of the two cylinders simplified the transfer of steam from one to the other. Several compound beam engines are preserved at the Science Museum and elsewhere.

Although there were only minor improvements to follow in the second half of the nineteenth century, the beam engine continued to be used, particularly for pumping water. The last one is thought to have been manufactured as late as 1919. The reasons for the popularity of the beam engine, despite its size, were its simplicity of manufacture and reliability. Although the many links and rods looked complicated, they were simple to make and easy to adjust. The relatively slow-moving parts lasted for years without wearing out. When the beam engine was finally ousted in the twentieth century – after 200 years of service in various forms – it is hardly surprising that hard-bitten engineers felt nostalgic towards it. Of all engines perhaps the beam engine had the most character.

DIRECT-ACTING ENGINES

This group might almost be called 'non-beam engines' and it embraces most of the engines which are usually visualized when the words 'steam engine' are used. The essentials are a cylinder in which a piston and piston rod reciprocate, the drive coming directly from this piston rod, usually via a connecting rod and crank, but with no beam.

One of the first direct-acting engines really was a beam engine without a beam, because existing beam engines could be converted. The vertical cylinder was retained with its piston rod protruding from the top. To guide the upper end of the piston rod along a straight line, it was connected to the mid-point of the middle link in a three-link mechanism. Such a linkage was earlier used by Watt and incorporated into his parallel motion, but this layout used a simple three-link mechanism fitted across the engine-house. From the mid-point a connecting rod ran upwards to a crank near the roof

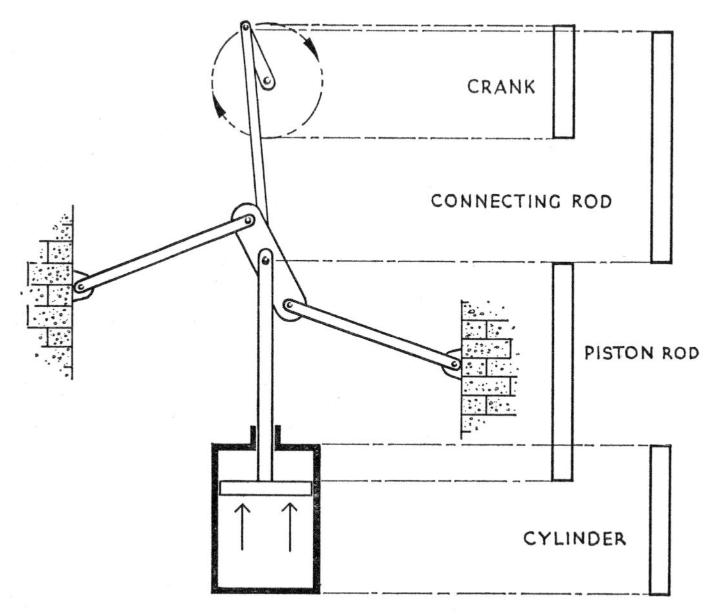

FIG 28 *Crowther's layout for an engine with a vertical cylinder and three-link straight line motion. Note the four tiers and the high drive shaft rotated by crank.*

rafters. Luckily beam engines had high engine-houses. This simple layout was designed in 1800 by Phineas Crowther of Newcastle-upon-Tyne. Many of his engines were installed at collieries in the Newcastle area for winding cages carrying men or coal up and down the shafts.

In contrast to Crowther's huge engines, a relatively tiny vertical engine was built in 1804 by Oliver Evans, the American co-inventor of the 'Grasshopper' beam engine. Evans's engine had a cylinder of only 6 inches diameter, yet it provided power for grinding plaster of Paris and sawing marble. He used a sliding crosshead to guide the end of his piston rod, but due to the limitations of his workshops this was made in hard wood.

Crowther's layout resulted not only in a very tall engine but also one in which the drive shaft was in a very high position. Although this arrangement was satisfactory for a winding engine it was not always convenient for other uses, and several engines were developed to overcome this problem of height. To compare these engines, consider the four 'tiers' of a vertical design; starting from the floor there is a cylinder, a piston rod, a connecting rod and, at the top, the crank with its flywheel and drive shaft.

Matthew Murray of Leeds designed a vertical engine in which the third and fourth tiers were combined in a 'hypocycloidal straight line motion'. This was an arrangement of gears to which the piston rod was attached directly. It guided the end of the piston rod along a straight line and it converted the reciprocating motion into a rotation. Although the engine was manufactured in about 1805, this type of gearing was still being used in certain instruments during the war of 1939–45, and it was also incorporated into many flat-bed printing machines.

Another vertical engine compressed into three tiers was Henry Maudslay's table engine patented in 1807. From the floor upwards its layout was crankshaft, cylinder and piston rod: the connecting rod was not eliminated, but it did not require a separate tier because it doubled back past the piston rod and cylinder. The cylinder was mounted on a table which was sometimes an ornate cast-iron structure. The upper end of the piston rod was attached to a cross-shaft which was guided along a straight line by wheels at each end running in parallel guide rails. Also attached to this cross-shaft, but outside the guide rails, were two connecting rods which ran downwards, past the cylinder, through slots in the table top to two cranks. By placing the cranks under the table a low position for the driving

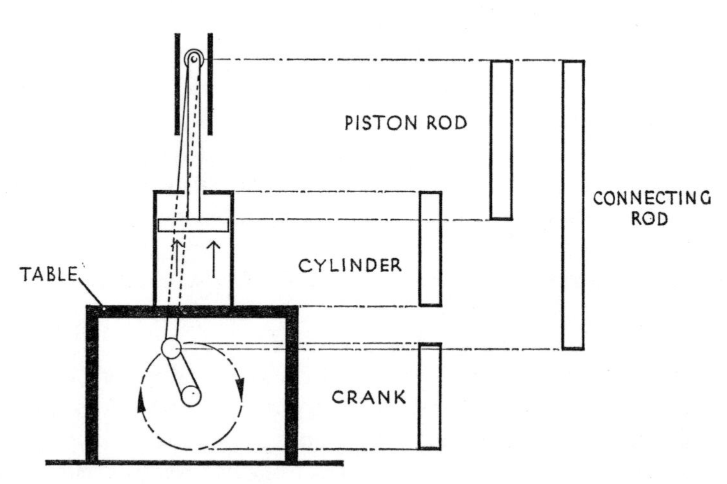

FIG 29 *The layout of a table engine, showing how space was saved and a low drive shaft obtained by the use of an overlapping connecting rod.*

PLATE 17 (*Top*) *Hancock's steam carriage* Automaton *of 1836 which could carry 22 passengers. The engine and boiler were at the rear of the vehicle.*

PLATE 18 *The 680-foot long* Great Eastern *of 1858: designed by Isambard Kingdom Brunel.*

PLATE 19 *The rotating 'runner' of a Pelton wheel shown before assembly. A jet of water was directed on to the centre of the twin buckets.*

PLATE 20 *Two hot-air engines. On the left Dr. Stirling's experimental model. On the right, a Robinson engine. Both engines are in the Royal Scottish Museum, Edinburgh.*

PLATE 21 *A model of a simple type of Babcock & Wilcox water-tube boiler. The fire was situated beneath the left-hand end of the inclined tubes.*

PLATE 22 *The Caledonian Railway locomotive No. 123 in the Glasgow Museum of Transport. This engine took part in London to Edinburgh races of 1888.*

PLATE 23 *A Willans high-speed steam engine. In effect this is three triple-expansion tandem engines built into one. It was built by Williams & Robinson Ltd. (now English Electric) in about 1900. Note the size of the chair.*

shaft was achieved, as well as an overall saving in height. This low drive and the compact layout were great advantages for some steam engine operators, and the table engine became very popular. Many workshops were powered by table engines, including Maudslay's own at Lambeth, where two were still providing power in 1900 when the works closed.

There were several other ingenious designs to reduce the overall height of vertical engines, including a 'steeple' engine, so called because its supporting framework was reminiscent of a church steeple.

Vertical cylinders were used by both Trevithick and Symington, and examples of both men's work can be seen in the Science Museum, South Kensington. Trevithick's stationary boiler and engine of about 1805 incorporated an arrangement of piston rod, connecting rods and cranks very similar to that used in the table engine. An interesting fact about these two engineers was their refusal to be dominated by traditional design. This is clearly demonstrated by their respective engines with horizontal cylinders, which were installed in Symington's steamboat *Charlotte Dundas* of 1801, as well as Trevithick's stationary engine of 1802, his steam carriage and at least one of his locomotives. Although the horizontal layout, with its reduced height, had several advantages, most designers feared that a piston moving horizontally would wear away the lower surfaces, due to the weight of the piston. A good oil would have reduced wear, but the science of lubrication was still almost unknown and, besides, there was no such problem with a vertical engine. None of these early horizontal engines remained in service long enough to settle the problem of wear.

Trevithick designed yet another type of direct-acting engine in 1812 – soon after his 'Cornish' beam engine. This was called a 'plunger-pole' engine and was originally designed to power water pumps, although later models were used for other purposes. The cylinder was in a vertical position, but instead of a piston and piston rod it had a cast-iron ram – the 'plunger-pole'. The action of this engine can be compared with two tin cans, one fitting inside the other; if the larger can is filled with water and the smaller one pressed into it, the water is forced out. Reversing this process, if water is forced into the outer can then the inner one will be moved out. In Trevithick's engine the inner can was the plunger-pole which was forced up by high-pressure steam instead of water. This involved sealing the space at the top between the two 'cans', but the

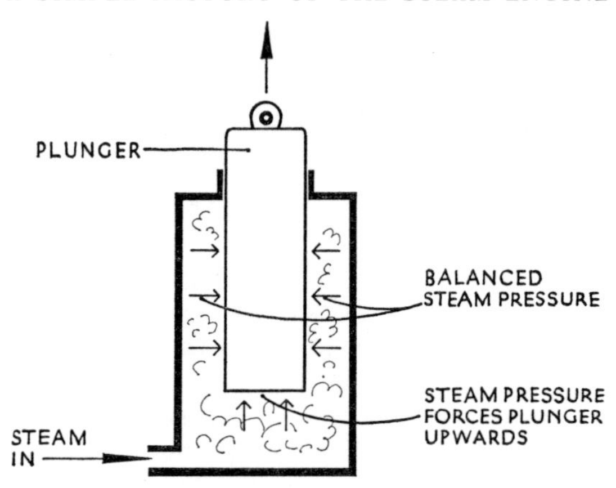

FIG 30 *How a plunger-pole engine works.*

design had one great advantage; the bore of the cylinder did not have to be accurately machined.

The plunger-pole was raised by steam at over 100 pounds per square inch, which was cut off after about one third of the stroke, the remainder of the stroke being powered by the expansion of the steam. When the plunger-pole reached the top of its stroke, its own weight returned it to the bottom again. If a water pump was being powered, it was placed directly below the engine, and rods from a crosshead on the plunger-pole ran down, one on either side of the cylinder, to the pump. Trevithick built several of these engines between 1812 and 1815 and they gave a very high duty. One or two were even added to Watt-type beam engines, converting them into compound engines.

In the 1820s a new machine-tool began to appear in workshops to supplement the lathes and boring machines. This was a planing machine which could produce a reasonably flat surface on metal, and once this was possible, a sliding crosshead could be used to guide the end of the piston rod. This development opened up a new range of direct-acting engines which could be used with their cylinders in almost any position.

The first commercially successful horizontal engines were produced in 1825 by Taylor and Martineau of London. Built for factory purposes, these engines had their cylinder, crosshead guides and crankshaft all supported on a cast-iron bed-plate. After a slow start this layout became universally popular, and even in the 1960s horizontal steam engines of this type were in regular use.

Vertical engines with a sliding crosshead had been tried as early as 1804 by Oliver Evans. However these early examples were handicapped by manufacturing difficulties and the height problem. Some solutions to the problem of height have already been described, but once the sliding crosshead was a practical proposition, further solutions were available. One of these was the 'diagonal' engine patented in 1822 by Marc Isambard Brunel, father of the more famous Isambard Kingdom Brunel. One or more cylinders were arranged side by side and inclined upwards towards a crankshaft fitted at any convenient height. Diagonal engines were used in textile and other mills, but perhaps they were better known as marine engines driving paddle-wheels.

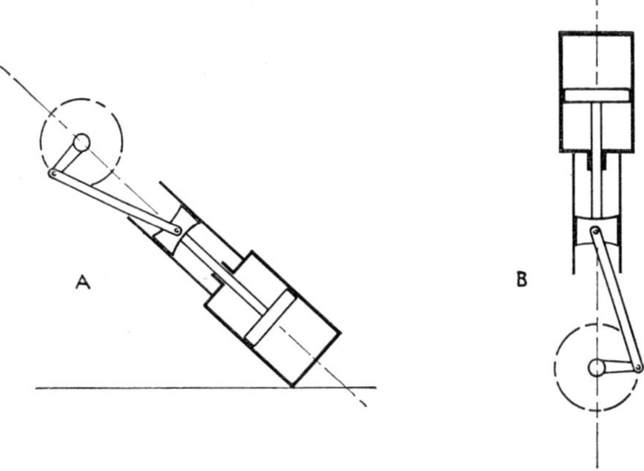

FIG 31 *A simplified diagonal steam engine and an inverted vertical layout.*

By inverting a vertical cylinder, the crankshaft was situated below the cylinder in a very low position, which was ideal for driving wheels or the propeller shaft of a ship. In the 1830s this layout was used by Walter Hancock for some of his steam road vehicles and in the 1860s it became popular for marine engines. Between these two dates James Nasmyth had used an inverted-vertical cylinder as a steam hammer, by attaching a hammer head directly to the lower end of the downward-pointing piston rod. Steam drove the piston and heavy hammer head upwards, then it was released and fell under gravity. Incidentally, James Nasmyth was the son of the artist who had painted a picture of Symington's first steamboat. Nasmyth, a very versatile genius whose inventions ranged from machine tools to telescopes, built many fine steam engines – which is not surprising as he served under Maudslay for several years.

For very many years the single-cylinder engine had been the backbone of steam power, but it had two disadvantages based on one geometric fact. When a piston is at either end of the cylinder, the piston rod, connecting rod and crank all lie along a straight line. In this position, whatever the pressure on the piston, it will not produce rotation until the crank is turned a few degrees off this 'dead-centre' position. The first disadvantage was that if an engine happened to stop with the piston at one of the dead-centre positions, it could not be restarted until it had been rotated by hand.

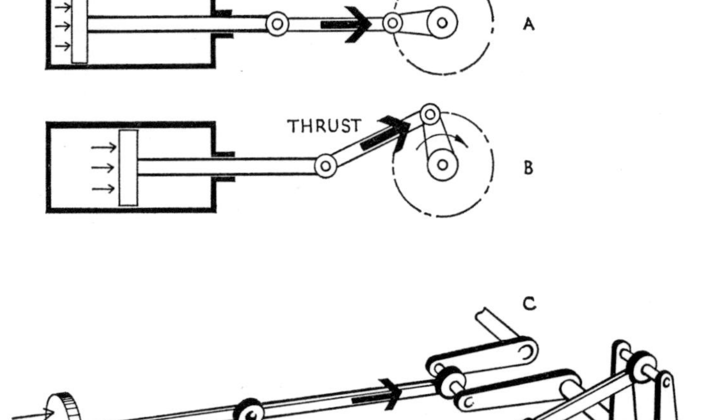

FIG 32 *The problem of top dead-centre and how it is overcome with a two-cylinder engine.* A *Piston, connecting rod and crank in the dead-centre position in which the thrust does not turn the crankshaft.* B *The connecting rod and thrust positioned for a large turning effort on the crankshaft.* C *A two-cylinder engine—only one piston can be in the dead-centre position at any time.*

As a crank rotates the maximum turning effect occurs about half-way through the stroke; thus, even when a constant pressure is applied to the piston throughout its stroke, the turning effect or 'torque' on the crankshaft varies from zero to a maximum and back to zero. This was the second difficulty because most of the machinery driven by steam required a steady torque. To 'iron out' these variations in the torque a large and heavy flywheel had to be fitted. A flywheel stores up energy during the maximum torque period and releases it as the crank passes the dead-centre position.

These difficulties were very largely overcome by designing an engine with two cylinders, which could be two simple cylinders or a compound pair. The joint crankshaft had two cranks, one for each cylinder, made at right angles to each other. By this arrangement, even if one piston was in the dead-centre position, the other was producing maximum torque. The starting difficulty was thus eliminated and a much more even torque was produced, resulting in a smaller flywheel.

Another variation of this twin-cylinder layout was called the 'non-dead-centre' engine. Only one crank was used, and the two piston rods were connected to this by a joint connecting rod shaped in the form of a large triangle. This arrangement not only kept the pistons out of phase but also enabled a more compact engine to be designed, because the connecting rod could double back in such a way that the crankshaft ran between the piston rods. Although the non-dead-centre engine was patented in 1839 by Elijah Galloway it was not widely used until the 1880s, when it became popular for mill and marine engines.

Early steamships and their engines are described in the following chapter, but two marine engines are described here, as they are important in the overall story of the steam engine. Marine engineers trying to fit large engines into ships were particularly worried by the length of the conventional steam engine, and so a variety of engines were designed to reduce this length. Some of these were merely geometric arrangements of links and rods, but the 'oscillating' engine and the 'trunk' engine were basically different in concept. Both of these were suggested during the Watt era and achieved prominence in the 1840s when built by John Penn, a London marine engineer.

A model engine with an oscillating, or rocking, cylinder was built by William Murdock and described in an earlier chapter. Comparing this engine with the conventional layout, it did not have a connecting rod, the piston rod being connected directly to the crank. As the crank rotated and the piston rod moved in a straight line the cylinder had to rock. Aaron Manby in 1822 and Joseph Maudslay in 1829 successfully powered vessels with oscillating engines, but it was not until John Penn revived the idea in 1838 that they were widely used. Penn introduced several improvements including more efficient steam supply valves – the supply of steam to a rocking cylinder was always a major problem with this design. Nevertheless Penn constructed many oscillating engines for paddle steamers, ranging from small river boats to ocean-going vessels.

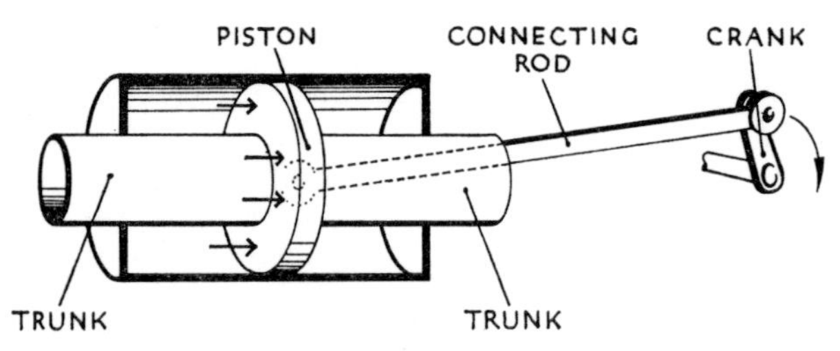

FIG 33 *The double-trunk engine with its unusual piston and no piston rod. Note that the area of the piston subjected to steam pressure is the same in each direction.*

Penn's other contribution was the 'double-trunk' engine which was developed from an idea patented by Watt in 1784. As with the oscillating engine, other versions had been tried, but Penn's was the best and he patented it in 1845. The 'double-trunk' engine was designed to drive a ship's propeller, for by this time the dominance of the paddle-wheel at sea was being challenged. Propeller shafts presented a number of difficulties because they were in a very low position within the hull of a vessel.

A horizontal engine was a natural choice if it could be fitted athwartships between the central propeller shaft and the side of the vessel. But horizontal engines were long, and Penn decided to eliminate the piston rod. He replaced it by a large hollow tube, or 'trunk', which carried the piston and emerged from each end of the cylinder (hence the name 'double-trunk'). The connecting rod ran from the crank inside this trunk to a pivot in line with the piston, and in this way the connecting rod was effectively connected to the piston. One important feature was preserved by this design – it was double-acting. There is no piston rod in a normal petrol engine, but this cannot be made double-acting. The double-trunk engine powered many merchant steamers and warships.

By 1850 the direct-acting steam engine had 'arrived'. As a result, the popularity of the beam engine declined, and in the second half of the century the term 'steam engine' would automatically imply one which was direct-acting.

CHAPTER 6

Steam-Powered Transport

The experimental era (1800–1825)

The improved steam engines of the nineteenth century gave those inventors who were striving to achieve steam-powered transport a new hope. For many years their efforts had been frustrated by unsuitable engines. Denis Papin had suggested a steam-engined boat a hundred years before one emerged; but he had no engine. Similarly, Jonathan Hulls of Gloucestershire designed a steam-powered vessel in 1736 and even patented it. But as the only engine available was the slow and cumbersome Newcomen-type, it is not surprising that there is no evidence of Hulls' boat ever having been developed beyond the design stage.

The inventors of the early 1800s were able to make rapid progress with working models and prototypes. These were followed by full-size road vehicles, ships and railway locomotives which actually worked. Then came the economically successful designs representing the final stage of invention. It is interesting to note that railway locomotives led in this final phase, despite the fact that steam road vehicles and ships had made their initial appearance many years before the first locomotive.

Cugnot's steam-engined road vehicle of 1769 has already been described, and it was followed by Murdock's model. This little vehicle also had three wheels and was powered by a miniature steam engine with a cylinder of only ¾ inch diameter. It was fitted with a fixed vertical cylinder and half a beam, because both cylinder and crank were close together on the same side of the pivot. Steam was supplied from a tiny boiler heated by the wick of a spirit lamp. Murdock demonstrated his model on several occasions, and one of these – so the story goes – took place in the churchyard of a Redruth church. The parson happened to see the model and, thinking it was the devil himself breathing fire, fled in panic from the churchyard.

Murdock experimented with his model steam 'carriage' for several years during his very limited spare time and by 1786 he had decided to patent his design. Boulton and Watt did not show any interest in the model, so Murdock acted on his own for once, and set out by coach from Cornwall to London. He never reached the Patent Office, however, for Boulton intercepted him near Exeter – whether by accident or design we do not know – and persuaded the loyal Murdock to return to Cornwall. On the way back they stopped in Truro and Murdock demonstrated his model to Boulton. It ran round in circles carrying a load of 'the fire-shovel, poker and tongs', but Boulton did not share Murdock's enthusiasm: in fact, he wrote to Watt asking him to send more beam engines for Murdock to assemble and thereby divert him from his steam carriage.

In 1801 Richard Trevithick built a full-size steam carriage and, according to reports, it climbed Beacon Hill in Camborne on Christmas Eve. A few days later it was tested on a turnpike road, the driver being Trevithick's cousin Andrew Vivian, but a rut in the road upset the steering and the carriage overturned. The passengers retired to an inn and enjoyed a good meal while the carriage caught fire and was destroyed.

A second steam carriage was built by Trevithick in 1803 and demonstrated in the streets of London. It had three wheels, the single front wheel being steerable while the two 8-foot rear wheels were driven by the engine. The boiler and engine were between these large wheels and the carriage was mounted above the engine, making the whole vehicle rather high and clumsy. While travelling at 5 or 6 miles an hour under the control of John Vivian (the son of Andrew) the steering again gave trouble and the carriage crashed. Trevithick presumably decided that the roads were not suitable for steam carriages, for he now turned his attention to railways.

While Trevithick was experimenting on the roads, progress was also being made with steam-powered ships, for the size of engines and boilers was less of a handicap on water than on the narrow roads. One problem which had to be overcome, however, was that of converting engine power into forward movement. A propeller or even paddle-wheels may seem an obvious solution to us, but for centuries ships and boats had been propelled by the power of the wind in sails or the muscles of men working oars and paddles. It is not surprising, therefore, to find that one of the first attempts to power a boat with a steam engine also used mechanical paddles. Six of these were positioned vertically down each side of the boat and moved by a single-cylinder double-acting atmospheric engine. The designer of this ingenious vessel was an American engineer called John Fitch and during 1787 he made several runs on the Delaware river, travelling at 3 miles per hour. A few years later Fitch made another steamboat which also used separate paddles – this time three were mounted over the stern. This boat, the *Experiment*, carried fare-paying passengers in 1790, but unfortunately Fitch's company was a financial failure.

Another American designer produced a steamboat in 1787; he was James Rumsey and his boat was jet-propelled. Incidentally, the jet was water, not the hot gases we associate with modern jet engines. The idea of employing a steam engine to pump water through a nozzle at the stern of a boat was not new, for it had been patented as early as 1729 by an English inventor, Dr. John Allen. Rumsey's boat made two trips on the River Potomac after which the project was abandoned.

The early pioneers of steam navigation, including Fitch, Rumsey and several others, were finding it very difficult to proceed from the 'prototype' to the 'working' stage. The five stages of successful invention previously mentioned were briefly: scientific research, plans and ideas, models and prototypes, practical working examples and, finally, the economic success. The prototypes might be successful experiments, but they could not perform any really useful work.

One of the first steamboats capable of useful work was the *Charlotte Dundas*, built in Scotland during 1801. Several people contributed to the success of this project and the supporters of each claim the glory exclusively for their favourite. However, it all started with an Edinburgh banker, called Patrick Miller, who made several experimental boats propelled by paddle-wheels. This was not an original idea, and the boats did not have a mechanical source of power: up to thirty men were employed to turn capstans which drove the paddle-wheels, and even horses were tried. One of the men was James Taylor, the tutor of Miller's sons, who did not take to this manual work and therefore suggested to Miller that they should install a steam engine. (Opinions differ as to who suggested the idea, but this is really not important because Fitch and others had already made steamboats by the year 1787.) Taylor did make an important contribution by introducing Miller to one of his friends, William Symington, who was an experienced steam engineer.

Symington's father had helped to install the Boulton and Watt engines at Torryburn and Wanlockhead, which have already been mentioned. Young William was destined for the church, but the attraction of the steam engine proved too great and he became an engineer at the lead mines in the neighbouring villages of Leadhills and Wanlockhead. In 1786 Symington made a model steam carriage which he demonstrated at the University of Edinburgh. Further experiments with road transport were dropped when he was asked by Miller to provide an engine for a paddle-boat. Although Symington had first received his steam engine experience with Boulton and Watt beam engines used for pumping purposes, he soon started designing and building his own engines, some of which were installed at the lead mines.

Symington agreed to design an engine for one of Miller's boats and in his own words:—

> 'Upon this mutual understanding I proceeded to erect a small steam-engine upon the principle for which I had previously procured a patent, having two cylinders of four inches diameter, each making an eighteen inches stroke. This engine having been constructed by my direction and under my eye, I caused it to be fitted on board a double-keeled vessel then lying upon a piece of water near the house of Dalswinton; and this being done, an experiment was made in the autumn of the year 1788, in presence of Mr. Miller and various other respectable persons, and the boat was propelled in a manner that gave such satisfaction that it was immediately determined to commence another experiment upon a more extended scale.'

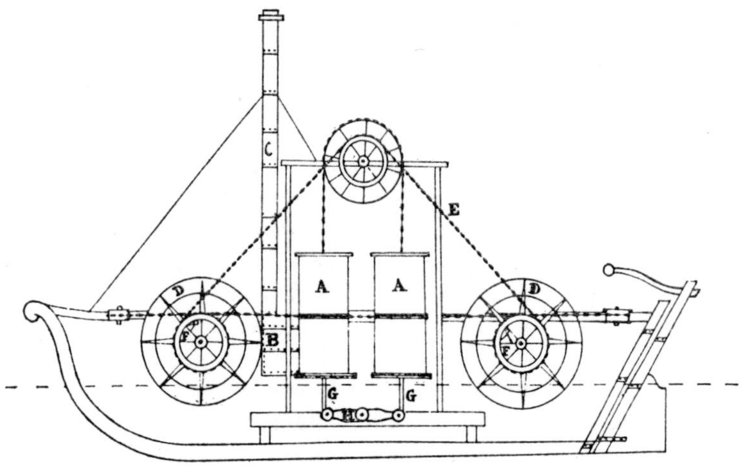

FIG 34 *A drawing of Symington's first steamboat from his own book* Steam Navigation. *The two cylinders are marked* A, *the boiler* B *and the two paddle-wheels* D. *These paddle-wheels are mounted between the boat's two hulls.*

The 'respectable persons' included Robert Burns the poet, and Alexander Nasmyth, a well-known artist who painted a water-colour picture of the boat. This picture shows a 25-foot boat with two hulls – a catamaran – and two paddle-wheels fitted between the hulls in a tandem arrangement. The paddle-wheels were rotated by chains and a ratchet device, although the crank had been patented eight years earlier. Symington's engine was a Newcomen-type of atmospheric engine – for Watt's patent was still in force. This engine is preserved in the Science Museum, South Kensington.

A second boat was converted to steam; this time it was 60 feet long and two 18-inch cylinders were installed. In October 1789 the boat was ready for a trial on the canal which linked the Rivers Forth and Clyde. The boat set out at a steady 5 miles per hour but, according to Taylor, they had trouble with the paddle-wheels. Trouble or not, Miller withdrew his financial support and Symington returned to the lead mines and his stationary steam engines.

Eleven years later Lord Dundas, who was a proprietor of the Forth and Clyde Canal, asked Symington to design a steamboat capable of towing barges. Symington's own description tells the story:—

> 'A steam-engine was erected with a cylinder of double power, 22 inches in diameter, and making a four feet stroke, and fitted into a boat adapted to the power of the engine; and after making various experiments, I, in March 1803, took on board of the boat at Lock No. 20 of the canal, the late Lord Dundas, my patron, Archibald Spiers, Esq. of Edderslee, and several gentlemen of their acquaintance, and made the steamboat take in drag two loaded vessels, the *Active* and *Euphemia* of Grangemouth, Gow and Espline, masters, each vessel upwards of seventy tons burden, and with great ease they were carried, without the assistance of any horses, through the summit level of this canal to Port-Dundas, a distance of nineteen and a half miles, in six hours, although it blew so strong a breeze right ahead during the whole course of the day, that no other vessel in the canal attempted to move to windward; and this experiment not only satisfied me, but every person who witnessed it, of the utility of steam navigation.'

This boat was named *Charlotte Dundas* after the daughter of Lord Dundas.

The engine of the *Charlotte Dundas* was a great improvement over Symington's earlier engines, for several reasons. It had a separate condenser and air pump; but perhaps of more importance was the

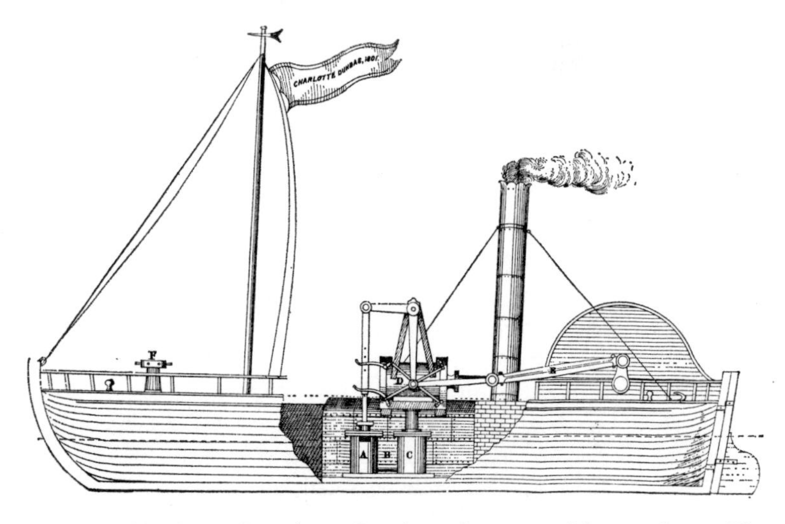

FIG 35 Charlotte Dundas—*Symington's successful steamboat. The horizontal cylinder is marked* D, *the boiler* B, *the condenser* C *and the air pump* A. *The engine drives a single paddle-wheel mounted in a slot at the stern.*

driving mechanism to the single paddle-wheel at the stern, because this eliminated a beam, and the cylinder was placed horizontally. The piston rod from the double-acting piston was guided in slides and the connecting rod linked this piston rod directly to a crank which drove the paddle-wheel. An overhung crank was devised by Symington which consisted of half a complete crank. This arrangement was patented by Symington in 1801.

Lord Dundas introduced Symington to the man who had led the canal-building era, the Duke of Bridgewater, who was so impressed with the reports of the *Charlotte Dundas* that he ordered eight steamboats. Unfortunately the Duke died in 1803 and the canal proprietors cancelled the order because they were afraid the wash from a steamboat would dislodge the banks of the canal.

Symington now abandoned his hopes of steam navigation and returned to mine engines. But his efforts were not completely wasted because *Charlotte Dundas* was examined by Robert Fulton and Henry Bell, two advocates of steam power who were to be more successful than Symington with a commercial steamboat.

Some five years after the *Charlotte Dundas*, another practical steamboat was built, but this one made a profit. The 'invention' of the steamboat was reaching its final stage. The designer of this vessel was an American named Robert Fulton, and it is interesting to note that the engine he fitted was bought from Messrs. Boulton, Watt and Co. Even though James Watt had retired seven years earlier, a vertical cylinder was still used in conjunction with a variation of the almost traditional beam. Instead of being straight the beam was L-shaped, with a pivot at the corner of the 'L'; then one arm was connected to the piston rod and the other to the paddle-wheels. Because a similar device was often installed to link the rods and wires which operated mechanical door-bells, this arrangement was called a bell-crank. The paddle-wheels were situated amidships, one on either side of the vessel. Side paddle-wheels proved to be a very popular layout which survived until recent years, being particularly suitable for the pleasure steamers which operated from Britain's holiday resorts.

So far, the name of Fulton's steamboat has not been mentioned, because this poses a problem. For many years she was thought to be called the *Clermont*, but researchers found she was originally registered as the *North River Steamboat*. In the following year the entry read the *North River Steamboat of Clermont*. Perhaps this was later shortened to *Clermont*. Anyway, she made her first commercial voyage on 4 September 1807 and continued to carry passengers on New York's Hudson river until mid-November. Modified during the winter, she reappeared to carry more passengers and remained in service until 1814.

Fulton ran into a certain amount of the trouble which beset many other pioneers – sabotage from operators of older equipment. He covered the paddles to prevent his passengers from being splashed, but these covers were supported by heavy frames, strong enough to withstand 'accidental' collisions with sailing boats. Fulton went on to build further successful steamboats and by 1812 he had built four, including one on the Mississippi.

In this same year the first commercial steamboat began operations in Britain. This was the *Comet* designed by Henry Bell of Helensburgh, the Clydeside resort. Bell's design was no great step forward and it required a number of modifications before the boat was ready to carry passengers. A passenger service between Glasgow and Greenock did not show a profit, and Bell decided to operate special cruises around the coasts of Scotland. These were more popular and aroused a great amount of interest in steam navigation. Some historians claim that Bell 'invented' the steamship while others have decried his efforts, but he was, at least, a good advertising agent for steam-powered boats. The engine from the *Comet* is preserved in the Science Museum, South Kensington.

By 1804 steam engines had made their debut on the roads and on the water; now it was the turn of the railways. Many years before the invention of the steam engine, railway tracks were born as an offshoot from the road. In Roman times a road surface was made up of flat slabs of stone arranged in a 'crazy paving' pattern. As wheeled carts and trucks became more widely used, someone decided that a full-width road was not necessary: two strips of flat stones, one under each wheel, would be sufficient. Then followed an even cheaper track, consisting of wooden planks under each wheel, but tracks had one great disadvantage – wheels ran off narrow planks all too easily. The answer to this problem was to fit flanges (a raised lip) to either the track or the wheels. A raised flange along the inside edge of each track would ensure that the wheels remained on the plank of wood. A better solution was to fit a flange to the inside face of each wheel. Iron replaced wood and for some time there was rivalry between the flanged plate, later called a 'plate tramway', and the flanged wheel. A glance at any railway wheel today reveals which proved to be more satisfactory.

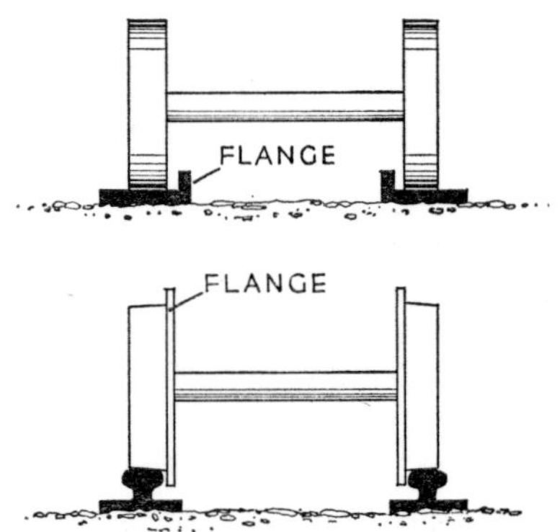

FIG 36 *Rail cross sections.* (*Top*) *A flanged plate tramway.* (*Bottom*) *Edge rails with flanged wheels.*

Tramways or wagon-ways linking collieries to the nearest canal, river or sea were widely used in Britain even during the early days of the Industrial Revolution. On these tracks several wagons loaded with coal were coupled together and drawn by a horse, but with increasing coal production the wooden tracks wore out very quickly.

Eventually cast iron became available and rapidly replaced wood. At this point steam power was considered, and it was first introduced by using a stationary beam engine to haul wagons up an incline by means of a long rope or cable. This arrangement was quite satisfactory for short distances, but a mobile locomotive would have much wider application.

Richard Trevithick has already been mentioned in connection with steam road vehicles and high-pressure stationary engines. His compact boilers and cylinders gave him a great advantage over any rival locomotive designers, and in 1804 his first locomotive was built for the Pen-y-darren ironworks in South Wales. It is reputed to have pulled five wagons loaded with ten tons of iron and seventy men; however it broke the cast-iron plate tramway in a number of places. Rather than re-lay the track with stronger plates the owners reverted to horse power.

Trevithick had certainly built the first successful prototype locomotive in the world, but whether it constituted a 'working' engine is questionable. Nevertheless, the Pen-y-darren locomotive was very advanced in design, for it included a high-pressure boiler, a horizontal cylinder and a sliding crosshead (i.e. no rocking beam). In addition, Trevithick released exhaust steam into the flue of the boiler, thereby increasing the draught – another invention about which there is much controversy. Many of these controversial claims grew into feuds long after the deaths of the inventors. With a shortage of real evidence, a son often made an extravagant claim on behalf of his father and forged documents were even produced.

Trevithick's second locomotive was built at Gateshead for use at Wylam Colliery in 1805. This village of Wylam near Newcastle later played an important part in railway history, but before describing these events, there were two other interesting developments – one in London and the other in Leeds. In 1808 Trevithick designed his third locomotive and demonstrated it near Euston Square in London. The locomotive was named *Catch-me-who-can* and it ran around a circular track enclosed by a high fence. For one shilling the public could see the novelty of a 10-ton locomotive running at 12 miles per hour. Unfortunately, attendances were disappointingly low, and when the locomotive crashed due to a broken rail, it was not repaired. Trevithick then devoted his attention to raising sunken ships.

Despite Trevithick's demonstrations many engineers were doubtful about the ability of a driving wheel to grip on a smooth metal rail. In 1812 John Blenkinsop received the first of several locomotives he had ordered from Messrs. Fenton, Murray and Wood, for use at Middleton Colliery, near Leeds. He went to great lengths to eliminate this fear of slipping wheels by fitting an extra pair of wheels with teeth, purely to drive the locomotive. These teeth engaged in the corresponding teeth of racks laid alongside each rail. This rack and pinion certainly eliminated slipping – in fact it is still used for steep mountain railways – but at the time its use did not spread because it was excessively costly to lay down and expensive to maintain.

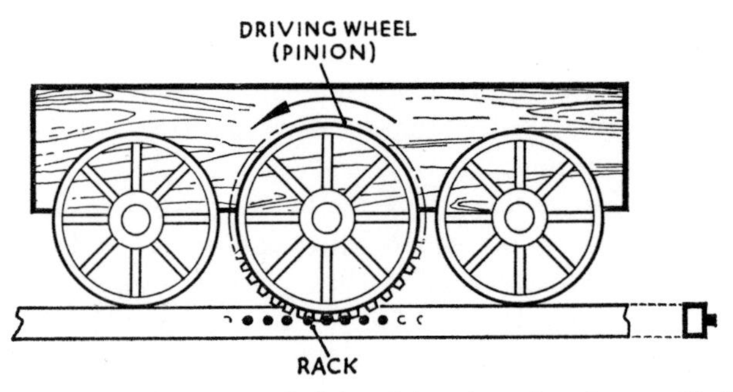

FIG 37 *Blenkinsop's rack and pinion drive, introduced to avoid slipping wheels.*

The main stream of railway development followed from Trevithick's second locomotive which had been ordered by Christopher Blackett, the owner of Wylam Colliery. There is no record of this locomotive working at Wylam, which may have been due to the fact that the locomotive gave trouble; or the fault may have been the weakness of the wooden wagon-way. Horses were retained to pull trucks from the colliery to a suitable place on the River Tyne where boats, known as keels, could be loaded with a cargo of coal – a distance of about five miles. In about 1808 the wooden way was lifted and replaced by cast-iron plates. After this alteration each horse could pull two trucks instead of one, but even this improvement did not satisfy Blackett, as the war against Napoleon had resulted in a serious increase in the cost of horses and their food. Oxen were tried, but they were too slow. Blackett was convinced a steam locomotive was the answer and he asked Trevithick to make another. Trevithick, however, declined the order. Undaunted, Blackett gave the task to one of his own bright young men, William Hedley, who was a 'Viewer' at Wylam Colliery.

In October 1812 Hedley carried out a series of experiments using a manually worked 'locomotive' with smooth wheels running on the cast-iron plates. He tried different weights loaded on to the 'locomotive' and in each case he added more trucks until the wheels skidded round. These tests convinced Hedley that a rack and cog wheel were not necessary and he went ahead with his first locomotive. In his own words:—

> 'An engine was then constructed, the boiler was of cast iron, the tube containing the fire went longitudinally through the boiler into the chimney. The engine had one cylinder and a fly wheel; it went badly, the obvious defect being want of steam. Another engine was then constructed, the boiler was of malleable iron, the tube containing the fire was enlarged, and in place of passing directly through the boiler into the chimney, it was made to return again through the boiler into the chimney, now at the same end of the boiler as the fire-place. This was a most important improvement. The engine was placed upon four wheels, and went well; a short time after it commenced, it regularly drew eight loaded coal waggons after it, at the rate of from four to five miles per hour, on Wylam Railroad, which was in a very bad state.'

(Extract from one of Hedley's letters published in 1882.)

The second and successful engine commenced work in May 1813, to be followed shortly afterwards by a twin. This famous pair were *Puffing Billy* and *Wylam Dilly* and they were to pull coal trucks on the Wylam railway until the colliery closed in 1862.

William Hedley was not a great inventor in the sense that Trevithick was. The latter was capable of great, original inventions, but he rarely carried them through to commercial success, whereas Hedley was a practical and determined engineer backed by an understanding proprietor in Blackett.

Hedley designed the Wylam locomotives incorporating what he considered to be the best features of his predecessors' designs in addition to some of his own ideas. He fitted two vertical cylinders which were connected to the beams of a 'grasshopper' type of beam engine (i.e. one half of a normal beam). A crank and gear wheels transmitted the drive to all four wheels, but the weight of the locomotive proved too great for the cast-iron plate rails and they frequently broke. In 1815 the locomotives were converted to eight wheels in order to reduce the load on each wheel. This layout remained until 1830, when the Wylam line was relaid with edge rails and the locomotives were each fitted with four flanged wheels.

FIG 38 *The* Wylam Dilly *locomotive powering a tugboat on the River Tyne, during a strike in 1822.*

Wylam Dilly proved its versatility in 1822 when it became a marine engine for a short period. During a strike by the men who worked the coal-carrying boats on the Tyne, the wheels of *Wylam Dilly* were removed and it was mounted on a boat to turn paddle-wheels. This improvised tug-boat successfully pulled a train of keels up and down the Tyne, despite some violent opposition from the keelmen. When the strike was over *Wylam Dilly* returned to the railway. Today this locomotive is displayed in the Royal Scottish Museum, Edinburgh, and *Puffing Billy* is in the Science Museum, South Kensington.

Wylam village was the scene of another important event in railway history – the birth of George Stephenson. When George was born in 1781, his father was fireman at Wylam Colliery and they lived in a cottage alongside the wagon-way. Young George also became a fireman and later a brakesman operating steam pumping and winding engines at various collieries. His progress was rapid, and in 1812 his employers, known as the Grand Allies, gave him control of all their colliery machinery. By this time Stephenson had moved to Killingworth, a mining village some five miles north of Newcastle, and he had a son Robert.

Stephenson was primarily concerned with stationary engines but, undoubtedly, he followed the progress of Hedley's locomotives at

Wylam. Eventually he persuaded the Grand Allies to let him build a locomotive for the Killingworth wagon-way. In July 1814 the *Blucher* was ready – named after the famous Prussian General who, the following year, fought against Napoleon at Waterloo. In some respects this locomotive was less efficient than Hedley's; for instance, it reverted to a single-flue boiler which produced less steam than the return-flue design used by Trevithick and Hedley. But Stephenson's design incorporated several important features: his connecting rods, from a sliding crosshead, drove directly on to the wheels, which were flanged and ran on iron edge-rails – a great improvement over the plates.

The fact that Stephenson's first locomotive was not of revolutionary design is of little consequence; of more importance was the fact that many other locomotives followed it. Trevithick had built prototypes, Hedley had produced working locomotives, and Stephenson made the railways a widespread commercial success – the subject of a later chapter.

CHAPTER 7

Raising Steam

The development of boilers up to 1830

'The engine had one cylinder and a fly wheel; it went badly, the obvious defect being want of steam.' These words of William Hedley were quoted in the previous chapter and reveal the important fact that, however well-designed an engine might have been, it was helpless without an adequate steam supply. And steam comes from boilers.

A good boiler not only supplies sufficient steam to keep the piston moving, but also is economical on fuel. With a steam engine the overall efficiency depends on two stages, first making the steam and then using it. If either of these stages is inefficient, then the overall efficiency suffers.

Although boilers have been mentioned from time to time in the story so far, they have not been described in detail or compared. This is a suitable place to enlarge upon boilers because, in the 1820s, they were almost more important than the cylinder and piston – hence Hedley's remark. One of the principal reasons for the importance of the boiler at this time was the fact that steam engines were being fitted to road vehicles, trains and ships; a large, heavy boiler was very awkward to mount on wheels and sometimes took up valuable passenger and cargo space. If a stationary engine required more steam, another boiler or two would be built nearby without any great detriment – except in coal consumption. More efficient boilers were obviously required for steam-powered transport.

What is a boiler and how does it work? A kettle boiling merrily over an open fire is the most elementary form of boiler, supplying a steady flow of steam from its spout. This steam is not, however, at a high pressure because it is escaping directly into the air, but if the lid of the kettle is fixed in position and a bung placed in the spout, then, as steam is formed, so the pressure inside rises. More steam increases the pressure still further, until the kettle explodes: so this is definitely not an experiment to be repeated.

The natural shape to contain a high pressure is a sphere, as demonstrated by an ordinary balloon, provided it has not been preformed into an odd shape. The alternative is a cylinder, preferably with domed ends. Thomas Savery used both these shapes for the high-pressure boilers he incorporated into his steam pumps of 1698 and onwards (see Chapter 2). As these boilers, made in beaten copper, were sometimes expected to withstand pressures of over 40 pounds per square inch, it is hardly surprising that there were several explosions.

Newcomen's engines required steam at normal atmospheric pressure – their power being derived from condensing the steam. This simplified the boiler problem and Newcomen reverted to something more like a kettle. But it was rather a peculiar shape, as can be seen from a diagram. The circular base of the lower cylindrical part was placed directly over the fire. This lower section was made in copper and the hemispherical domed top was usually of lead. Bricks were built up around the sides and top of the boiler as an insulator to keep the heat in. By building the side wall in line with the top of the boiler, an annular space was left around the smaller cylindrical base. This was made into a flue, called a 'wheel draught', through which smoke and hot flue gases passed before disappearing up the chimney.

Sitting in front of a blazing coal fire on a cold winter's night is one way to keep warm, but despite the heat given out, it is a very inefficient form of heating because most of the heat is wasted up the chimney. One of the main aims of boiler designers throughout the centuries has been to make use of these hot flue gases in addition to the direct heat from the flames of a fire.

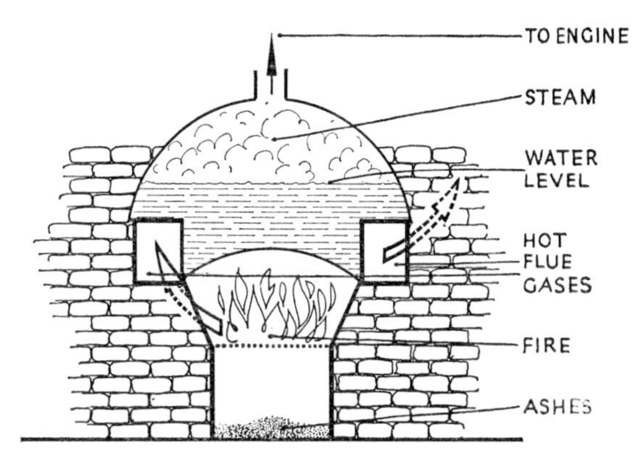

FIG 39A *An early circular boiler, Newcomen's haystack design.*

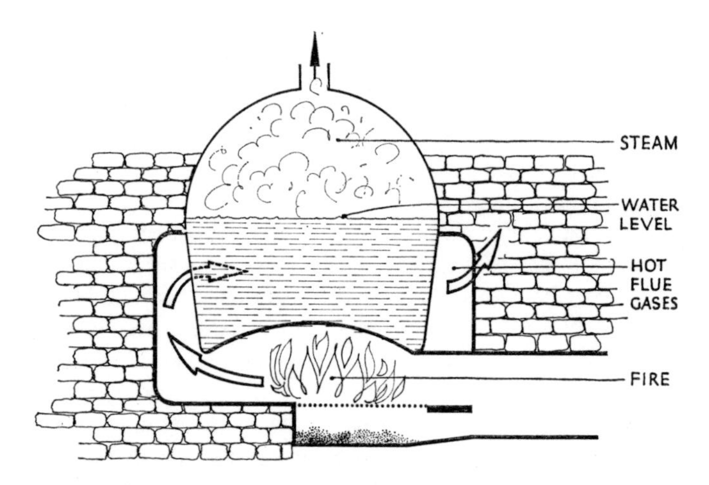

FIG 39B *An alternative circular-boiler—the 'tun'.*

Newcomen heated the bottom of his boiler by flames, and the sides by hot smoke and flue gases. This boiler with its distinctive domed top was known by a number of descriptive names, depending on the locality. 'Haystack' was perhaps the most popular, but 'beehive', 'balloon' and 'flange' were also used. The water level inside the haystack had to be maintained above the seam joining the lead dome to the copper base, otherwise the joint overheated and failed. One engine burned out four boilers in four years – probably due to carelessness in the control of the water level.

In about 1725, hammered wrought-iron plates, riveted together, were introduced in place of copper and lead – a change which resulted in more robust boilers. Another alteration favoured by some engineers, including John Smeaton, was the discarding of the step around the side of the boiler. The wheel draught idea was retained, but three of the flue's four sides had to be built in brickwork – whereas Newcomen's flue had two in brick while the other two were the actual boiler. This was called a 'tun' boiler because its shape was similar to a brewer's fermenting vat or 'tun'. Although the tun boiler was slightly less efficient than the haystack, it was more reliable and easier to manufacture.

If very large quantities of steam were required, the engineers could increase the size of the domed boiler, but only up to a certain point, beyond which considerations of strength together with manufacturing difficulties became too much of a problem. One solution was to install several small boilers. When his engines required large boilers James Watt discarded the circular tun and used an elongated 'wagon' boiler. In shape this resembled an American covered wagon, or a rectangular box with half a cylinder as a lid. An end view of the wagon boiler was very similar in shape to the tun, but its length might be two or three times its width. The fire underneath and wheel draught around the sides were retained.

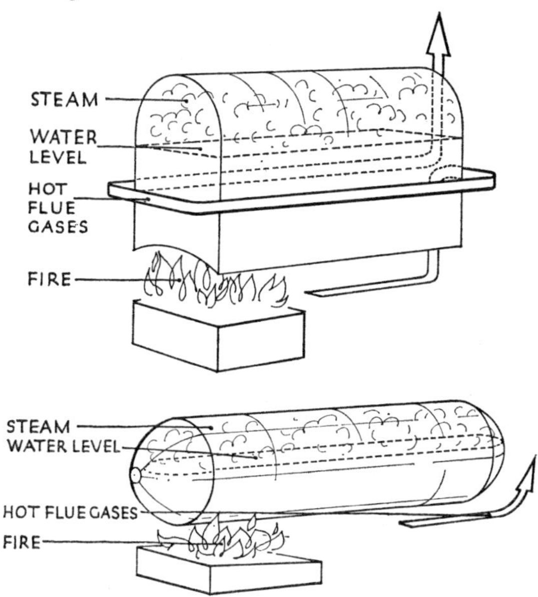

FIG 40 *Two elongated boilers. Note: the brick flues are omitted and the flow of hot flue gases are shown by an arrow. (Top) The wagon boiler favoured by James Watt. (Bottom) A cylindrical boiler with domed ends. Several different flues could be used.*

The wagon boiler remained in use for half a century, despite one major handicap – it could not produce high-pressure steam. At the pressures used by Watt it was perfectly satisfactory, but if pressures over about 5 pounds per square inch were produced the flat sides of the wagon would bulge precariously. By adding stays across the boiler linking the two sides, this bulging was reduced. Even with strengthening stays, the wagon boiler was of no use to men like Richard Trevithick and Oliver Evans who visualized steam pressures of over 100. They both came to the conclusion that a cylindrical boiler was the best compromise between the wagon and a sphere.

Although a spherical boiler was the ideal shape to withstand high pressures, it was difficult to make and it had only a small heating surface – even smaller than the haystack and the tun. A long cylinder containing water, with a fire underneath, made a good boiler, but designers were for ever trying to increase the heating surface and not waste heat up the chimney.

Incidentally, it is almost impossible to say who invented some of the types of boiler, because water boilers had been used long before the steam engine came along and the 'new' boilers were often modified versions of existing designs. For example, Watt did not 'invent' the wagon boiler in the accepted sense (as is sometimes claimed), but he did improve it and give it prominence, which might be called the final stage in the process of invention.

Trevithick and Evans further complicated the process of invention by developing similar boilers, at the same time, on opposite sides of the Atlantic. They both decided that more heat could be extracted from a fire, or furnace, by placing it inside the cylindrical boiler instead of underneath it. To do this, they fitted a large tube through the boiler from end to end and placed the fire inside the tube. With water all around the fire, a much larger heating surface could be obtained, and so it became a more efficient boiler. The fire was placed at one end of the 'fire tube' or 'furnace tube' and the hot flue gases from this fire passed along the tube – giving up some of their heat to the surrounding water – before escaping up the chimney.

The next stage, to extract even more heat from the flue gases, was to direct them back along the outside of the cylindrical water tank in brick-built flues. Both Trevithick and Evans developed this improvement and, in the case of Trevithick, it became the famous 'Cornish' boiler already mentioned in Chapter 5.

From this point the ideas of the two engineers diverged. Evans reversed the flow of gases by putting his fire underneath the boiler and carrying the flue gases along the bottom before they passed through a flue pipe inside the boiler below water level. It has been suggested that he did this, not to improve efficiency, but to reduce the risk of an explosion if a careless operator let the water level drop below the safety mark which, in the earlier layout, was above the fire tube.

So far the boilers described were primarily for stationary engines and widespread use was made of brickwork for the flues, but brick flues were not practical for the boilers of road vehicles, locomotives

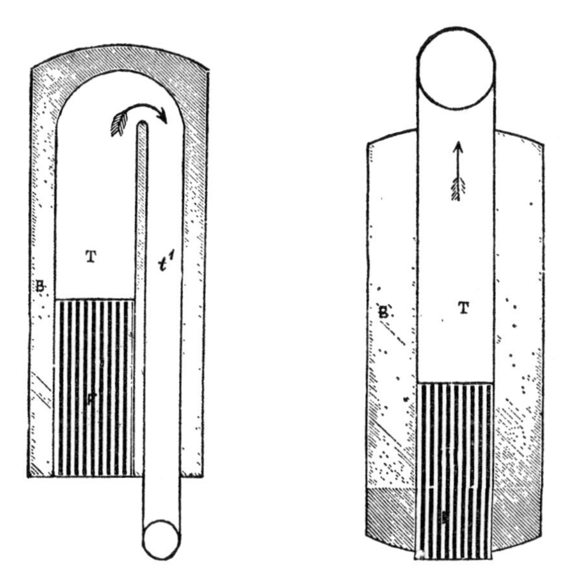

FIG 41 *Two variants of the fire-tube boiler. (Left) Hedley's return-flue. (Right) Stephenson's straight through tube.*

and steamboats. Of course a cylindrical boiler with a fire underneath it could be used and, even better, a boiler with a fire-tube, yet even this did not satisfy Trevithick. He devised a return-flue boiler for his locomotives. This had its fire inside a tube which carried the flue gases through the water to the far end of the boiler, but, before emerging, the tube made a U-turn and returned to the fire end, still surrounded by water. This obviously gave a great increase in the heating surface and led to a more efficient boiler. The return flue was later used very successfully by William Hedley and others.

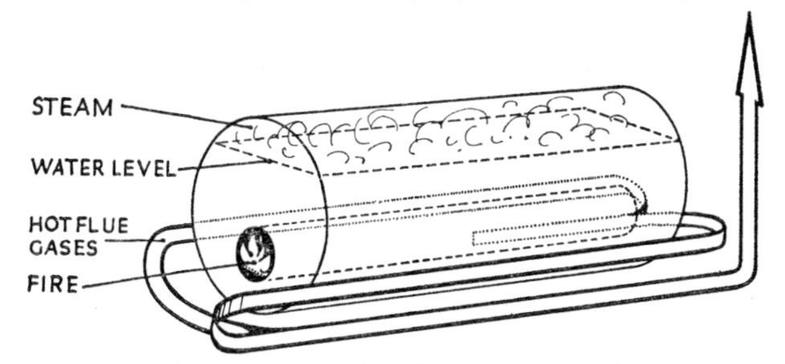

FIG 42 *A simplified drawing of a Cornish boiler showing the path of the hot flue gases.*

Two other ideas introduced by Trevithick were taken up by later locomotive designers. One was a method of drawing more air over the fire to increase its heat. This 'forced draught' was obtained by feeding waste steam from the cylinders into the chimney so that the hot steam rose upwards and drew air with it. The second idea was to mount the cylinder inside the boiler, thereby making use of the hot water to keep the cylinder hot, thus eliminating the need for a steam jacket around the cylinder.

Fire-tube boilers, as pioneered by Trevithick and Evans, developed along several paths, one starting from the Cornish and another from the return flue. The Cornish boiler survived in its own right, but it was augmented by a similar design with two fire tubes instead of one. This was called a 'Lancashire' boiler and was patented in 1844 by William Fairbairn and John Hetherington of Manchester, although it may have been used in Cornwall ten years earlier. Even as late as 1913, almost half the stationary boilers in the United Kingdom were either Cornish or Lancashire, such was their success.

A whole range of boilers was developed from the simple fire-tube layout in an effort to produce a compact and portable boiler. The basic idea of heating water by both fire and hot flue gases was retained, but a great improvement was introduced to extract even more heat from the gases. Instead of feeding these gases through one large tube, surrounded by water, they were fed through a large number of small tubes. This was called a multitubular boiler, and once again it was invented independently by two great railway engineers. Marc Séguin patented his design in 1828 and later applied it to a locomotive on the Lyons and St. Étienne Railway in France. Robert Stephenson and Henry Booth brought out a similar boiler the following year and installed it in the famous *Rocket* locomotive.

The multitubular boiler as used in most locomotives had a straight-through layout with the fire at one end and chimney at the other. A return flue idea, using multitubes, was introduced many years later for marine work, and this became known simply as a marine boiler or alternatively a 'Scotch' boiler. On land one of the most successful designs with a return multitubular flue was the vertical Cochran boiler.

So much for boilers based on the simple kettle principle of heating a tank containing water until steam is produced. Another method of heating water is sometimes incorporated into the hot water system of a house, and this consists of passing water through tubes which are

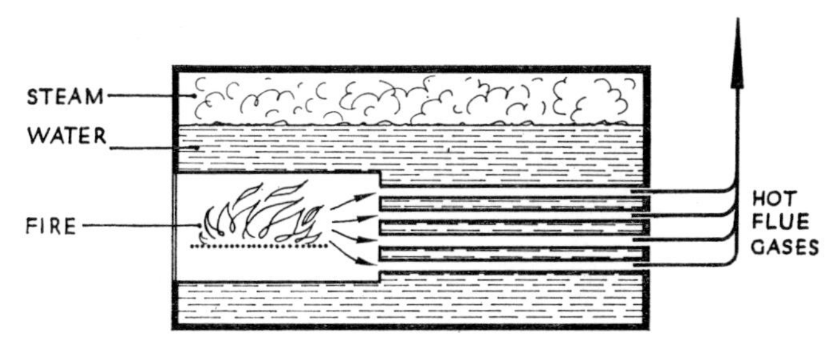

FIG 43 *A multitubular type of boiler. The hot flue gases divide and pass through a large number of tubes. The locomotive boiler is a development of this design.*

heated externally. Most gas- and oil-fired central heating boilers work on this principle, as does the Ascot type of water heater. This layout is called a water-tube boiler, and it has the great advantage of being able to produce hot water or steam very rapidly.

The water-tube idea for heating water is not new; in fact one such device was found at Pompeii, where it had been buried during the famous volcanic eruption of A.D. 79. While heating water was relatively easy, producing high-pressure steam presented more of a problem.

One of the first pressurized water-tube boilers was built by William Blakey in about 1770 in conjunction with his ideas for an improved version of Savery's steam pump. His idea for a boiler was a good one, but the seams of his copper tubes could not stand up to the steam pressure. James Rumsey, the designer of a jet-propelled steamboat, visited England in 1788 and patented several designs for boilers, including some with water tubes. Unfortunately, Rumsey died before these designs could be developed.

Arthur Woolf required high-pressure steam for his compound steam engines and he did actually produce water-tube boilers following his patent of 1803. He used large cast-iron pipes which were quite strong enough, but he had trouble with leaks caused by the pipes expanding and damaging their connections. The pipes were arranged side by side, some heated by the fire and others by the hot flue gases. Additional pipes running across the water pipes collected the steam. It may be remembered that Woolf's partner Humphrey Edwards went to France and built compound engines there. Edwards modified Woolf's boiler by reducing the number of tubes to three and making them larger – thus making its classification as a 'tank' or 'water-tube' boiler rather difficult. This version became known as a 'French' or 'elephant' boiler and it was widely used on the Continent where the 'Lancashire' boiler never gained popularity.

Several water-tube boilers followed from Woolf's and these were particularly popular with the designers of steam-powered road

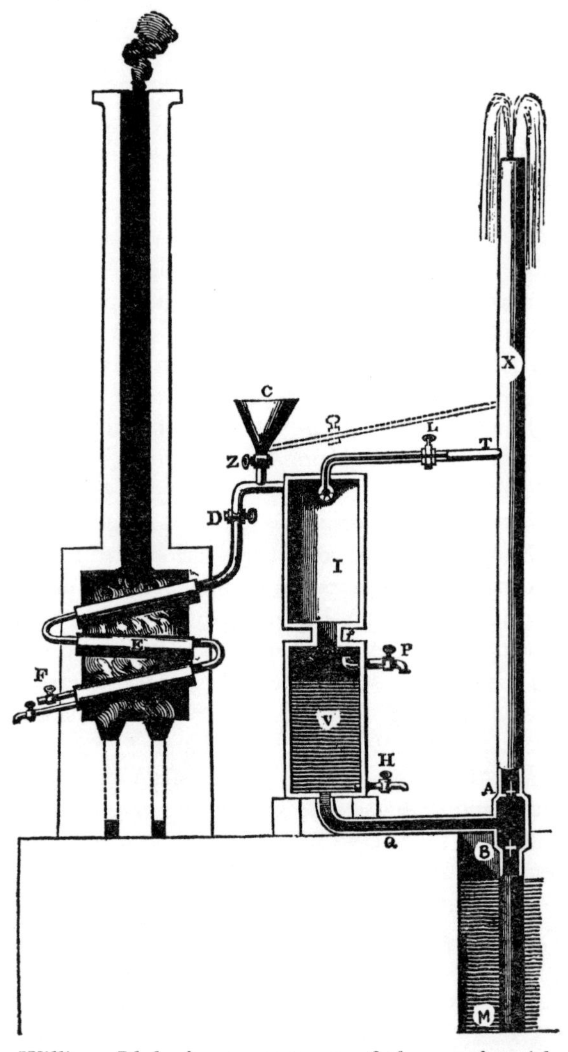

FIG 44 *William Blakey's steam pump of the 1770's with a simple water-tube boiler marked* E. *The pump is a development of Savery's design.*

vehicles who required quick-steaming boilers. These vehicles of the 1830s are described in a later chapter. Although water-tube boilers were a limited success, it was not until the end of the century that designs by Stirling, Yarrow, and Babcock and Wilcox really challenged the supremacy of the 'Lancashire' boiler.

Another method of producing steam quickly is to spray water on to a red-hot metallic surface, and if this is carried out in an enclosed space high-pressure steam can be generated immediately. This idea of a 'flash' boiler was suggested in 1736 by John Payne, yet it was not tried out until about 1823 when Jacob Perkins started experimenting. Perkins pumped hot water through square hollow bars made from cast-iron and kept at red heat by a furnace. Not only did this produce instant steam, but it heated the steam to a very high temperature. Later this became known as 'superheated' steam.

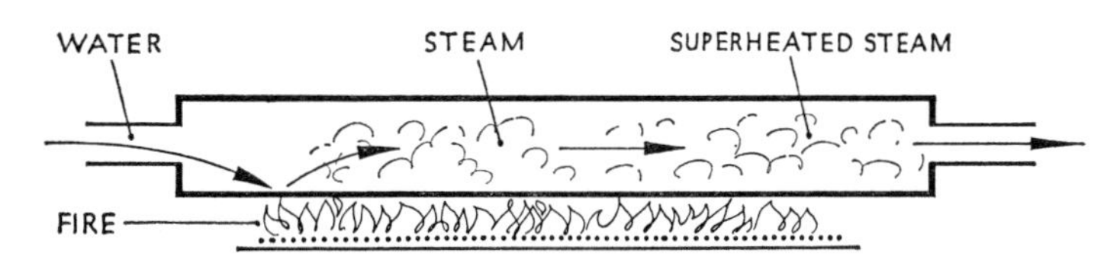

FIG 45 *The principle of the 'flash' boiler which makes instant steam.*

Perkins invented and built an experimental engine in 1827 which was powered by high-pressure steam. Because of the difficulty in making valves to control steam at high temperatures and pressures, Perkins eliminated exhaust valves by allowing the steam to escape through holes, or 'ports', in the cylinder wall. During the early part of its stroke the piston covered these ports, but just before the end of each stroke they were uncovered and the waste steam escaped. Although this type of engine was not developed at this stage, it was 're-invented' later and called a 'Uniflow' engine.

The pressures Perkins could produce, which reached some 1,500 p.s.i., were far higher than anyone could use for practical purposes in the 1820s. However, Perkins himself had a novel idea, according to the 1829 Supplement of Stuart's *Descriptive History of the Steam Engine*.

'Some time ago Mr. Perkins made some experiments on a steam-gun before the Duke of Wellington, in which the high-pressure steam, raised in his safe generators, was employed. At a distance of thirty-five yards the bullets penetrated eleven deal planks, each one inch thick, at a distance of one inch from each other. The steam employed was equal to 900 pounds on the inch. In some of his experiments the gun-barrel moved on a joint, and the balls being supplied from a sort of funnel or hopper, as rapidly as they were shot off (about 250 per minute, and in one experiment at the rate of nearly 1000 per minute), the stream of shots could be directed in the same manner as the spout of a fire-engine. A deal plank was pierced with a line of holes from one end to the other. One may easily judge of the effect of such an engine directed against a line of soldiers.'

CHAPTER 8

The Railway Age

1825–1850

Within the past few years the mighty steam locomotive with its distinctive smells and noises has become part of history. Like so many giants, both natural and mechanical, it has succumbed to a more efficient and versatile rival. Yet, even with electric motors and diesel engines instead of steam power, railway systems in many parts of the world are, themselves, fighting to survive against the challenge of the motor car and the aeroplane. It is a far cry from the boom years of the nineteenth century, when the alternative to a train was either a stage coach or a canal barge.

The beginning of the railway age was an exciting period of history, but the date when it began is open to argument. Claims to have operated 'the first railway' abound, and strangely enough most of these claims are true. The difficulty arises over the definition of a railway. If it is merely a 'vehicle on rails', then the first is lost in history, but if a public railway is considered then the *Edinburgh Encyclopaedia* of 1825 suggests:—

> 'The first public railway company is understood to have been instituted at Loughborough, in the year 1789, where the late eminent Mr. Jessop had the merit of first employing the edge-rail.'

This is rather a vague claim, but the early use of an edge-rail instead of the usual plate-rail is interesting.

A more positive 'first public railway' was the Surrey Iron Railway, which was authorized by Parliament to carry goods on payment of a fee. It was opened in 1803 and consisted of horse-drawn trucks running between Wandsworth and Croydon.

To most people a horse-drawn truck on rails does not constitute a railway; it must be mechanized. Steam power came in 1804 with Trevithick's locomotive, but as this ran on a private line it was not available to the public. Of course Trevithick's *Catch-me-who-can* locomotive of 1808 carried fare-paying passengers – round and round in circles. This was followed by a number of successful colliery locomotives all of which ran on private lines.

In 1821 a public railway between Stockton and Darlington was authorized, and in the following year George Stephenson was appointed as engineer. He negotiated for an Act of Parliament giving the company power to carry passengers as well as goods, and to use steam locomotives. With his son Robert and some friends, Stephenson set up the firm of Robert Stephenson and Co. to build locomotives. Although Robert's name was used, he was abroad from 1824 to 1827 and only contributed to the early stages of this important project.

The first locomotive was ordered in 1824 from the Newcastle workshops of Robert Stephenson and Co. and was a development of George Stephenson's colliery locomotives. The opening date of the new line was fixed for 27 September 1825, and during the previous evening a trial run was made, carrying the directors of the railway company as passengers. George Stephenson accompanied the distinguished passengers while his brother James drove the engine, which was later called *Locomotion*. The Stockton and Darlington line thus became the first public railway to use steam power.

Locomotion was not a great step forward in design and it is estimated to have developed about 10 horse power, which enabled it to pull a load of about 60 tons at a speed of 5 miles per hour. *Locomotion* was withdrawn from the line in 1841 and was later used as a stationary engine. In 1857 it was put on display at Darlington station, but on several subsequent occasions this famous locomotive has run under its own steam. One of these was the centenary of the opening of the Stockton and Darlington line.

When the Stockton and Darlington Railway opened, it possessed one passenger coach. The name 'coach' was very appropriate because it was the body of a stage coach mounted on the wheels of a wagon. For several years *Locomotion* and the locomotives which followed were fully occupied hauling coal wagons, leaving the passenger coaches to be pulled by horses. Incidentally, one horse could pull a fully loaded coach on the railway, whereas two, or even four, horses were normally used on the roads. The twelve-mile journey between Stockton and Darlington could be covered at a speed of about 10 miles per hour. The fare was 1s. 6d. inside the coach, or 1s. 0d. outside on the roof: the difference in comfort between first and second class was very marked in the early days. Horse-drawn coaches were withdrawn in 1833, but by this time a steam-powered passenger railway had been opened.

While Robert Stephenson was in South America he happened to

PLATE 24 *A large compound steam engine made by John Musgrave & Sons Ltd., of Bolton and used to produce electricity for Glasgow's trams in 1901. This 4,000 horse-power engine had one high-pressure cylinder (far right) and two low-pressure cylinders (centre and left).*

PLATE 25 *An atmospheric gas engine made in 1867 by Crossley Bros., Manchester to a Langen and Otto patent. The 'free' piston was shot upwards by an explosion of gas below it—thus creating a partial vacuum. Atmospheric pressure then forced the piston down for the working stroke. This ½ horse-power engine is in the Royal Scottish Museum, Edinburgh.*

PLATE 26 *A model of a large 'High Head' boiler as made by Babcock & Wilcox Ltd.*

PLATE 27
A 45-ton steam crane lifting a dry-dock gate at Leith Docks. This crane has recently been scrapped and replaced by a gantry for loading containers.

PLATE 28 *A 12-ton Marshall steam roller of 1928, in the process of being restored. It has a single cylinder engine with piston valves.*

meet Richard Trevithick, who had just been rescued from an alligator-infested river. Trevithick, having made and lost several fortunes, was penniless at the time and young Robert Stephenson gave him some money. When Robert Stephenson returned to England in November 1827 he found his father struggling to build a passenger and goods line between Liverpool and Manchester. This was a major undertaking covering thirty miles of difficult terrain including hills, which required cuttings or tunnels, and the notorious marsh Chat Moss. Because George Stephenson was so involved with the civil engineering works, his locomotive workshops were being neglected.

The leading locomotive engineer of the day was a rival, Timothy Hackworth, who had previously worked for William Hedley and George Stephenson himself. Hackworth's *Royal George*, built in 1827 for the Stockton and Darlington Railway, had vertical cylinders mounted in the inverted position above the rear wheels of the locomotive. With this layout the connecting rods could drive the wheels directly, thus making them much shorter than in the case of earlier locomotives. *Locomotion*, for example, had vertical cylinders with the piston rod emerging from the top of the cylinder and long connecting rods doubling back down to the wheels. Hackworth's simple layout was a great step forward, but he retained a parallel motion linkage to guide the crosshead.

Early in 1828 Robert Stephenson took charge of the Newcastle locomotive works and during the next four years he revolutionized locomotive design. Although he was aided by his father and others, including Henry Booth, Robert was the leading figure. One of the stipulations in the Act of Parliament which authorized the Liverpool and Manchester Railway was that any locomotives employed must consume their own smoke. The coal-burning locomotives of the Stockton and Darlington Railway or the colliery lines were slow and dirty, making them unsuitable for a passenger train.

George Stephenson and Henry Booth had experimented with a 'smokeless' boiler early in 1827 and Robert included this in his first new-style locomotive, the *Lancashire Witch*. Built for the Bolton and Leigh Railway, this locomotive started work in July 1828. It had a boiler with two fire tubes – so making it a forerunner of the 'Lancashire' boiler – and burned coke. Bellows under the grate overcame the difficulty of burning coke, and both fire tubes exhausted into a common chimney. The layout of the drive from the cylinders to the wheels was delightfully simple and introduced what later

H*

became standard practice; the piston rod was guided by a sliding crosshead, and a connecting rod linked this to a crankpin on the driving wheel. *Lancashire Witch* had two cylinders mounted above the rear wheels and inclined in such a way that they drove the forward wheels. The rear wheels were driven from the forward wheels by coupling rods.

Work on the line between Liverpool and Manchester was nearing completion early in 1829 when the Directors had second thoughts about steam locomotives. They were worried because it was a considerable advance from hauling coal trucks at a colliery to a regular service for passengers over thirty miles. Two independent engineers were appointed to investigate the possibilities, and they reported that stationary engines hauling trains by means of long ropes were the best solution to the problem. George Stephenson fought this suggestion with great vigour and produced figures to prove that stationary winding-engines would be almost twice as expensive to operate as his locomotives. The Directors decided to organize a trial to find the best locomotive and determine its performance before committing themselves. A prize of £500 was offered and conditions for entry were specified: the trial was to take place in October 1829.

Five engines appeared before the judges at Rainhill on the Liverpool and Manchester line, some nine miles from Liverpool. The trials lasted several days and, after great excitement, the prize was awarded to the now legendary *Rocket*. Amongst the defeated entries were the *Sans Pareil* made by Timothy Hackworth and the popular favourite, *Novelty*, by John Braithwaite and John Ericsson. The *Rocket*, entered by the Stephensons and Henry Booth, averaged 13·8 miles per hour with a load of almost 13 tons and, running light, it reached 29 miles per hour.

In appearance the *Rocket* was similar to the *Lancashire Witch*, except that it had small rear wheels which were not coupled to the front driving wheels. However, there was a less obvious alteration which was to revolutionize locomotive design. This was the introduction of a multitubular boiler – an event already described in the previous chapter. From 1830 onwards practically every locomotive built in Britain had a multitubular boiler working on the same basic principle as the one designed by Robert Stephenson and Henry Booth.

By fulfilling all the conditions laid down by the Directors of the Liverpool and Manchester Railway, the *Rocket* proved beyond doubt that a steam-powered locomotive was suitable for their railway. Both the *Rocket* and *Sans Pareil* were bought for the new line which was finally opened on 15 September 1830. In the meantime another railway opened and began the first regular steam-hauled passenger service in the world. This was the Canterbury and Whitstable Railway which opened in May 1830 using another Stephenson locomotive, the *Invicta*, for its six-mile route. This locomotive is preserved at Canterbury while the *Rocket* and *Sans Pareil* can be seen at the Science Museum, South Kensington.

Other railway companies sprang up in Britain, on the Continent and in America, and many of these used British locomotives in their early days. In 1828 a Stephenson locomotive was sent to France for the St. Étienne-Lyons Railway, but its single-flue boiler did not produce sufficient steam. The following year Marc Séguin built a locomotive with a multitubular boiler to work on this same line. Stephenson and Séguin's parallel invention of the multitubular boiler has already been described, but Séguin introduced an unusual method of increasing the draught through his fire tubes. He placed two huge rotating fans on the tender, where they were driven from the wheels, and these supplied air to the boiler through flexible leather pipes. A blast of steam in the chimney – used by Stephenson and others – proved to be a more satisfactory method of increasing the draught.

Another Stephenson locomotive of 1828 was exported, this time to America for the Delaware and Hudson Canal and Railroad Company, but there were delays in the building of the line. The first scheduled steam train to carry passengers in America was run by the South Carolina Canal and Rail-Road Company on Christmas Day 1830. Three years later they had a continuous route of 136 miles – the longest in the world.

Britain's first long-distance trunk line was opened in 1837 when the Grand Junction Railway linked the Liverpool and Manchester Railway to Birmingham. This enabled mail to reach Liverpool from London in the amazing time of 16½ hours, which included a horse-drawn coach journey over the first stage to Birmingham.

On 13 June 1842 the new and young Queen Victoria made her first railway journey when she travelled from Slough to Paddington – a daring event for a monarch, causing a considerable stir at the time. By 1848 the railway from London to Edinburgh and Glasgow was completed and Britain was in the midst of a period of 'railway mania'. Railways were being built at a fantastic rate, but building railways is not within the scope of a book about engines, although the two were closely linked in the early years. George Stephenson had built both railways and locomotives, and in his retirement he became the first President of a Mechanics' Institute which later developed into the Institution of Mechanical Engineers. Engineering had become too wide a subject for one man to cover and, from this period, engineers began to turn to either 'civil' or 'mechanical' engineering. The building of roads, canals and railways came under the heading of civil engineering, while the mechanical field covered steam engines, including locomotives, and machinery of a wide variety. Most of the established engineers of the first half of the nineteenth century were 'civils' who indulged in some mechanical work, but the next generation of engineers tended to concentrate on only one branch.

Following in his father's tradition, Robert Stephenson designed locomotives as well as supervising the building of railways, but his chief rivals of the 1830s, Timothy Hackworth and Edward Bury, were predominantly mechanical engineers. By a strange historical coincidence, each of the three designed a similar locomotive in about 1830, all three used horizontal cylinders mounted in a new position, between the lines of wheels, and later called 'inside' cylinders. Hackworth's *Globe* had twin cylinders under the driver's platform, while the two more important engines, Stephenson's *Planet* and Bury's *Liverpool*, had their cylinders between the front wheels driving on to a crank in the rear axle.

Stephenson included a refinement which had been described to him by Trevithick some time earlier. The twin cylinders were placed in the flue – or to be more precise in the 'smoke box' – below the chimney. In this position the smoke and hot gases heated the cylinder walls before passing up the chimney, thus eliminating the need for a steamjacket and resulting in a considerable increase in power.

Locomotives incorporating this same layout of twin cylinders between the wheels were being built as recently as 1948, but outside cylinders – similar to the *Rocket*'s – remained more popular through the years, particularly for large engines. Perhaps it should be mentioned here that in the complex business of designing locomotives there was rarely a sharply defined right or wrong solution to any problem. A number of equally successful locomotives could have been designed, all of which might have been completely different because a designer had such a wide range of choice. He could vary the position, size and number of cylinders, the type of valves and their operating mechanism, the number of wheels and their arrangement, or boiler details such as fuel, tubes and draught producer – and a multitude of other technical and even artistic details. This variety resulted in many arguments and some compromises, but most of these are too complex to describe in this book. Even a simple matter of engine power resulted in a difference of opinion. Some railway companies favoured large locomotives to haul their heavy trains while others preferred to employ two smaller locomotives 'double-heading'.

Although Hackworth designed the *Globe* with horizontal cylinders, most of his designs retained the 'old fashioned' vertical or steeply inclined cylinders. These were quite satisfactory for the slow-speed coal trains of the Stockton and Darlington Railway but, by 1837, the demand for higher speeds forced Hackworth – and one or two other advocates of vertical cylinders – to abandon them in favour of horizontal cylinders.

The first locomotives with horizontal cylinders outside the front wheels – later to become the traditional position – were built in 1834 by George Forrester of Liverpool. Strangely enough, this layout was not an immediate success because the wide spacing of the cylinders, in conjunction with a short wheelbase, resulted in a peculiar swaying motion. This movement was so pronounced that these locomotives were nicknamed 'boxers'. Two years later an extra pair of wheels was added to lengthen the wheelbase and thus counteract the unsteadiness.

Trains at this time travelled at between 20 and 25 miles per hour, but in 1835 that great and versatile engineer Isambard Kingdom Brunel caused consternation when he announced his plans for the Great Western Railway between London and Bristol. He was going to place the rails 7 feet apart, instead of using the widely accepted 'gauge' of 4 feet 8½ inches, and he envisaged speeds of up to 40 miles per hour. On this project Brunel acted chiefly as a civil engineer – building the track – while various makers supplied locomotives, usually to Brunel's specification. Some of the resulting engines were freaks, but the two star performers were appropriately called *North Star* and *Morning Star*. These fine locomotives came from the workshops of Robert Stephenson and Co. and were not built to Brunel's specification. When the first stage of the new line was completed it was the *North Star* which hauled the inaugural train from Paddington to Maidenhead. On this run an average speed of about

28 miles per hour was maintained and, after a grand luncheon party, the return journey was completed at an average of 33½ miles per hour. Later the *North Star* pulled a load of 45 tons at the impressive speed of 38½ miles per hour.

The success of the two Stephenson locomotives was even more pronounced in view of the failure of the others: the position was serious. Brunel's young assistant, a mechanical engineer from the North East called Daniel Gooch, was appointed to the position of Locomotive Superintendent, and his task was to provide some practical locomotives. Gooch designed a new range of locomotives based on the *North Star* and a hundred were ordered for delivery before 1843. Many manufacturers were involved in this first large-scale order of standardized locomotives. It was a great engineering feat, because drawings, specifications and templates had to be supplied to each maker in order to ensure that parts could be interchanged between locomotives – 'interchangeability' as it is now called. Sixty-two of the hundred consisted of passenger expresses, and these were ordered from seven manufacturers, including the workshops of Matthew Murray and James Nasmyth, both of whom were mentioned earlier – Murray in connection with Blenkinsop's rack-railway of 1812 and Nasmyth as the pioneer of the steam hammer.

Although the original *North Star* remained in service until 1870 it was eventually scrapped. However in 1925 a full-size reproduction was built at the Swindon works of the G.W.R. to take part in the Stockton and Darlington Centenary celebrations. This new *North Star* is now in the Swindon Railway Museum.

Robert Stephenson was a remarkably prolific designer, and in 1841 he introduced his 'long-boiler' to provide more steam and give a greater efficiency. Normal boilers were about 8 feet long, but Stephenson increased this to over 11 feet, thus increasing the length of the tubes in the multitubular section. Naturally this gave a very much greater heating surface between the hot flue gases and the water, with a consequent increase in efficiency. Unfortunately, the early locomotives with a short wheelbase tended to be unsteady at speeds over 35 miles per hour. With an increased wheelbase 'long-boiler' locomotives were successful and many were built, some with inside cylinders and, later, others with outside cylinders.

Locomotives of this period burnt coke in order to reduce the nuisance of smoke, yet most engineers would have preferred to burn coal because it gave out more heat. In 1841 Samuel Hall of Nottingham built a small locomotive called the *Bee* which could burn coal without producing too much smoke. Hall's scheme was to provide plenty of air and deflect the smoke back over the fire. Unfortunately the boiler over-heated very easily and coal, as a fuel, did not supersede coke until the 1850s.

In the late 1840s the Great Western Railway's workshops at Swindon were producing not only many of the best locomotives but also some of the best designers. Two of Daniel Gooch's pupils became leading engineers of their day; Archibald Sturrock was a sound, practical mechanical engineer like Gooch, while the other was an imaginative – almost eccentric – inventor who turned his attention to many things besides locomotives. This colourful character was called Thomas Russell Crampton and many of his locomotives included unusual features – some successful, others not. One of the features of Crampton's designs was the large pair of driving wheels fitted at the very rear of the locomotive, whereas most designers placed their driving wheels nearer the centre. Although Crampton engines were not an outstanding success in Britain, they became popular on the Continent and by 1864 nearly 300 had been built in France and Germany; many were still in service on French railways in the early years of the present century.

Robert Stephenson built some of his 'long-boiler' engines with large rear driving wheels, but their axles were forward of the firebox in contrast to Crampton's which were even further back under the driver's platform. One of Stephenson's engines, with rear driving wheels, took part in the famous trials of 1845–46 which were organized by the Government to help them decide between the two gauges of rail in general use. It was obviously impracticable for part of the country to adopt the 7-foot broad-gauge of the Great Western Railway while the remainder used 4 feet 8½ inches. Unfortunately, the trials did not provide a clear-cut answer. Gooch's G.W.R. locomotives created a very favourable impression on the observers because they were the more powerful, but Stephenson's long-boiler engines were more economical. Other factors had to be considered by the Government, however, including the mileage of track already laid in the rival gauges. The outcome was a decision in favour of the narrower gauge, and gradually the broad-gauge disappeared – although in some places it lasted until 1892.

Some remarkable locomotives were built in the late 1840s. One of these had three cylinders – two small ones outside the wheels and a large one inside. This layout, patented by George Stephenson and

William Howe in 1846, was an attempt to overcome the unsteadiness of the 'long-boiler' engine, but only two were built. Another experimental engine was the *Cornwall*, designed by Francis Trevithick, son of the famous Richard Trevithick. The boiler of this engine was placed below the axle of the huge driving wheels in an attempt to keep the centre of gravity as low as possible. As no great advantage was gained by this idea, *Cornwall* was rebuilt in 1858 as a conventional locomotive.

'Uniflow' and 'compound' steam engines have already been mentioned in previous chapters describing stationary engines. In 1849 an experimental locomotive with 'uniflow' cylinders was built by J. I. Cudworth. After three years in service it was discarded, probably because the large pistons, required to cover the steam ports in the cylinder walls, caused excessive swaying. As engines increased in size, and speeds rose, designers had to pay more attention to balancing. Even a wheel which has been balanced 'statically' can produce a violent vibration when rotated at a high speed – as can be proved on a motor car. But 'dynamic' balancing of the wheels for a locomotive is not the only problem, because the large pistons and rods reciprocating at high speeds also produce some unusual effects. For instance, a piston moving along a horizontal line can produce a bouncing action up and down on the rails. One of several engineers who tackled this problem was Crampton, but a more important contribution was made in 1845, when a balanced engine was designed by John G. Bodmer, a Swiss engineer who settled in Lancashire. By 1849 some work on the theory of balancing – a very complex subject – had been published by M. Le Chatelier.

A steam locomotive with a compound engine was patented in 1850 by James Samuel, though credit for the idea is usually attributed to his employee John Nicholson. Two locomotives were converted to the new 'continuous expansion' process. The method of operation was not exactly the same as a conventional compound engine, which expands steam in a number of cylinders, one after the other. Nicholson used the high-pressure steam in one cylinder, then, about halfway through the stroke, some of the steam was allowed into the second cylinder, thus using both cylinders at the same time. The idea was not a great success, and compound steam locomotives did not reappear for about twenty-five years.

A great contribution to steam engine design was made by several engineers who developed methods of controlling the steam supply to engine cylinders. Most locomotives were fitted with D-slide valves which were operated by eccentrics, as pioneered by William Murdock some fifty years earlier. This simple mechanism was quite satisfactory for stationary steam engines supplying power for a mill or factory, but locomotives had two extra requirements to fulfil: they had to be able to reverse and they had to provide a variable amount of power. Even with a specific load of trucks or carriages a locomotive requires more power to travel uphill than on the level, and it does not have a gearbox like a car. In most early locomotives the power was governed by a throttle or regulator valve which controlled the supply of steam – as a household tap controls the supply of water to a basin. Even though the throttle control was not very efficient it worked, and so the more pressing problem was for a reversing mechanism.

It may be remembered that some single-cylinder steam engines can turn in either direction, but reversing a two-cylinder engine involves altering the timing of the valves. All the early reversing valve gears required the complete disconnection of the eccentrics from the valve-operating spindles in order that the valves could be moved by hand and reconnected in their new position. This was a time-wasting procedure, and between 1835 and 1841 several improvements were made to speed the process, one of the most successful being a variation of Carmichael's gear which uses a control lever in conjunction with notches or 'gabs'. The real breakthrough came in 1842 when the 'Stephenson link motion', controlled by a simple lever, was introduced. Although named after the Company, this cunning mechanism was reputed to have been suggested by a draughtsman, William Williams, and brought to a practical form by a pattern-maker called William Howe. Robert Stephenson approved of the idea and incorporated it on a locomotive, and from that time onwards it remained a very popular reversing gear. But this was not the end of the story, for there was still the problem of controlling the steam supply.

In 1838 John Gray patented a mechanism which regulated the power by varying the timing of the steam cut-off point. As previously mentioned, Cornish beam engines achieved a high efficiency – but low power – by admitting steam into the cylinder during only the first part of the piston's stroke, which enabled the expansive power of steam to be used. By contrast, maximum power could be obtained by allowing steam into the cylinder throughout the stroke. The advantage of using the cut-off point as a means of control, instead of a simple throttle, was the increased efficiency under lighter loads.

This was very important because a locomotive might spend a considerable time running at reduced power, since travelling along a level track is far less arduous than starting from rest or climbing a hill. Full power would be reserved for this type of work – or for very high speeds.

Although it is probable that Stephenson's link motion was originally designed as a reversing gear only, it was easily adapted to incorporate steam control by a variable cut-off point. This dual purpose, combined with simplicity, convenience and economy of fuel led to its widespread adoption for a period of 100 years. Despite its success, some rival link motions were produced. For instance, when Daniel Gooch devised one for the Great Western Railway in 1843 it remained in use until 1888. Gooch's link motion is reputed to have been derived from Williams' original idea but, like Howe, he modified it into a practical form.

Another type of valve gear, introduced by E. Walschaërts of Belgium during 1844, turned out to be another 'delayed-action' invention, which became very successful on the Continent during the latter part of the century and in Britain even later. Many of the last great steam locomotives built in Britain incorporated the Walschaërts valve gear.

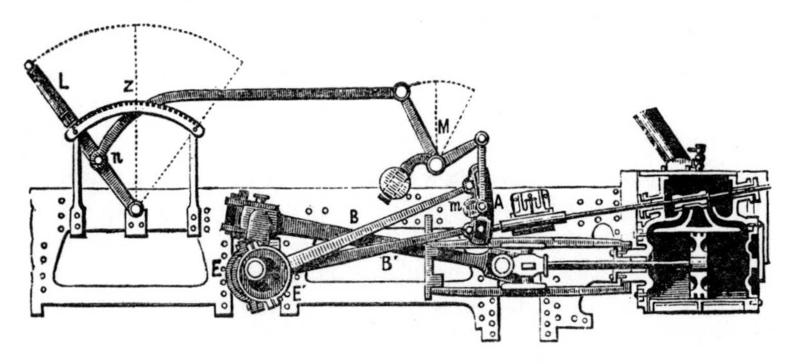

FIG 46 *The Stephenson link motion controlling the steam valves. The control lever* L *is shown in the position for maximum power in reverse, while the vertical position* Z *is 'neutral'. Movement beyond* Z *gives gradually increasing power in the forward direction. Maximum power is achieved by a late steam cut-off point.*

With the invention of Stephenson's link motion the railway locomotive had reached an important stage in its development. All the essentials were there; the inventions which followed tended to be refinements to increase the efficiency of the power unit and boiler, or geometrical changes involving the overall layout. Bogies and brakes for the wheels, large wheels versus small wheels coupled together, as well as other aspects of railway history are very interesting, but this book is trying to tell the story of the steam engine as distinct from the steam locomotive.

CHAPTER 9

The Working Steam Engine

Successes and failures in the mid-nineteenth century

'Judging from the rapid strides the Steam Engine has made during the last forty years, to become a universal first mover, and from the experience which has arisen from that extension; we feel convinced that every invention which diminishes its size, without impairing its power, brings it a step nearer to the assistance, of the "world's great labourers" – the husbandman and peasant, for whom as yet it performs but little. At present it is made occasionally to tread out the corn; – What honours await not that man who may yet direct its mighty power, to plough, to sow, to harrow, and to reap?' These words, probably written in 1823, brought to a close Stuart's *Descriptive History of the Steam Engine*. His ideas were proved to be correct before another forty years passed by: the steam engine did diminish in size and its 'mighty power' was directed to ploughing.

The development of the steam engine itself, its experimental use to power transport, and the railway boom have already been described. During the mid-ninteenth century there were several interesting applications of steam power, some of which were successful while others failed. In the 1820s muscle-power was still very widely used, particularly on the roads where the stage coach was at the height of its popularity. During 1829 horse-drawn omnibuses were introduced into England from France and their first service was between Paddington and the Bank in London. Twenty-two passengers were carried inside a wide vehicle which was drawn by three horses, side by side. Charging a shilling for the complete journey or sixpence half-way, the owner, John Shillibeer, is reputed to have made a profit of £100 in the first week.

The muscles of men were being harnessed for a number of new projects including cycles and airships. In 1818 the two-wheeled 'hobby-horse' made its appearance; this forerunner of the bicycle was propelled by the rider pushing with his feet, as on a child's scooter. In the air, ballooning had grown in popularity since the first human ascent in 1783, but some aeronauts, not content to drift with the wind, wished to power their craft and fly wherever they desired. Some very remarkable airships emerged, powered by men operating oars, paddles or propellers.

Even as early as 1809, Sir George Cayley, the 'Father of British Aeronautics', wrote:

> '... we shall be able to transport ourselves and families, and their goods and chattels, more securely by air than by water, and with a velocity of from 20 to 100 miles per hour.
>
> To produce this effect, it is only necessary to have a first mover (i.e. engine), which will generate more power in a given time, in proportion to its weight, than the animal system of muscles.
>
> The consumption of coal in a Boulton and Watt's steam engine is only about 5½ lb. per hour for the power of one horse. ...'

Cayley studied the problem in great detail and even considered an internal-combustion engine. In 1816 he proposed an airship powered by two propellers driven from a steam engine. Cayley's airship was never built, but in 1852 a French engineer called Henri Giffard built and flew the first successful airship. Powered by a 3 horse-power steam engine, this craft could travel at about 5½ miles per hour and on one occasion covered some seventeen miles. Giffard

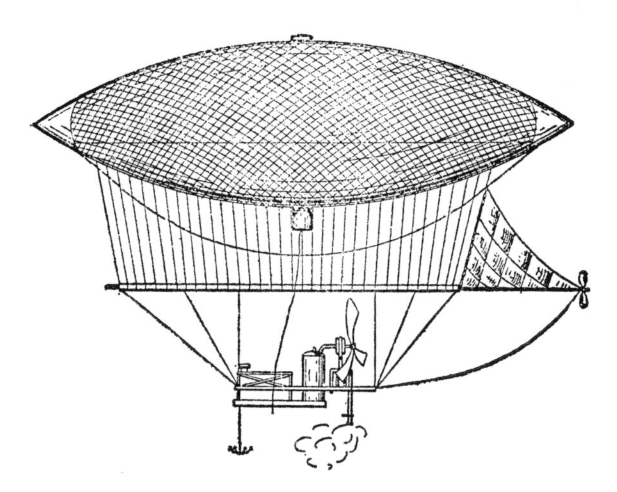

FIG 47 *A steam-engined airship built by Henri Giffard in 1852. The three horse-power engine drove an eleven-foot propeller which gave the airship a speed of just over 5 miles per hour.*

was not only an aeronautical engineer but also a steam engine designer of note and, in the mechanical engineering world, he is remembered as the inventor of the water injector for boilers. This device, produced by Giffard in 1859, was designed to replace water pumps as a method of forcing a supply of water into a boiler against the internal pressure. Within a few years most locomotives in Britain were using Giffard's injector.

The sheer bulk and weight of the steam engine was a great handicap, particularly when flying machines were being considered, and attempts to build an aeroplane with a steam engine never met with any real success. An ambitious design was patented in 1842 by William Samuel Henson for an 'Aerial Steam Carriage'. The proposed aeroplane had an enormous wing span of 150 feet and was powered by a steam engine of 25 to 30 horse-power driving two six-bladed propellers. Attempts to raise money to build a full-sized machine failed, despite some fanciful advertising using artist's impressions of the 'Aerial Steam Carriage' flying over London, the Channel and even the Pyramids.

Henson then decided to collaborate with his friend John Stringfellow on a less ambitious project – a flying model powered by a miniature steam engine. They built a 20-foot model which flew, but only in a 'powered glide', as its engine was not powerful enough to sustain even a level flight. Henson abandoned the project in 1847 and emigrated to the United States, leaving his friend to carry on with the experiments.

In 1848 Stringfellow made a new model with a wing span of 10 feet which he launched from an inclined wire. Historians differ on the achievements of this model, which is preserved in the Science Museum, London. However, it had an interesting miniature steam engine with a cylinder diameter of $\frac{3}{4}$ inch and a stroke of 2 inches. The horizontal engine was double-acting, and it was supplied with steam from a thin copper boiler heated by a naphtha lamp. Two propellers, 16 inches in diameter, were driven by the engine, and the complete model weighed under 9 pounds, including fuel and water.

Stringfellow abandoned his experiments for many years, but when the Aeronautical Society of Great Britain held its first exhibition at the Crystal Palace in 1868 he exhibited a model triplane together with a $\frac{1}{2}$ horse-power steam engine. The triplane was not a great success, but the steam engine was awarded a prize of £100 by the Aeronautical Society because it was the lightest engine in proportion to its power. Others followed Henson and Stringfellow's lead, yet the lightest steam engines were still too heavy for flying machines.

Small steam engines were also considered as a source of power for bicycles as early as 1818. A dated French print of that year shows a hobby-horse driven by steam, but it is unlikely that this machine was actually built. A number of drawings appeared during the following half-century depicting steam-powered bicycles and tricycles, but if these machines did exist, all trace of them has been lost. As the first commercially successful pedal bicycle was not produced until about 1865, the likelihood of a powered cycle before this date is rather remote. The bicycle mentioned was a French Michaux velocipede, or 'boneshaker' as it was called in Britain, and this preceded the more famous 'penny-farthing'.

In 1869, a year after Stringfellow's prize-winning engine, a French Perreaux steam engine was installed in a Michaux velocipede. This compact little engine was fitted beneath the saddle of the bicycle and drove the rear wheel by means of a round leather belt. Although this motor cycle was quite a practical vehicle it did not result in any production orders. Luckily it was not destroyed and is still preserved in France.

Steam-powered bicycles never developed into a practical means of transport, but steam coaches and omnibuses did appear on the roads. While the Stockton and Darlington Railway was still being built, several steam coaches were being designed and built for use on the roads. One of the most famous of these early designers was Sir Goldsworthy Gurney who, in 1825, patented his design for a coach propelled by legs pushing it along. Even though this was not a very practical device, Gurney's boiler was an improvement on other boilers for vehicles. By this time small steam engines of sufficient power were readily available; the main problem facing steam carriage designers was to produce a small boiler which could supply sufficient steam in a short time. Many variations were eventually produced, but they were generally water-tube boilers fired with coke.

Gurney's first coach, built in 1827, was powered by a two-cylinder engine which drove the rear wheels and pushing-legs, the latter being used to assist on hills. These were found to be unnecessary and later discarded. The body of this vehicle resembled a stage coach and could seat six passengers inside, with another twelve seats provided outside. A speed of 15 miles per hour was possible, but a stop had to be made every eight or ten miles to refill with water. Although contemporary prints show this coach on the London to

Bath road, it is believed to have operated principally in and around London over a period of two years.

Gurney built several coaches and, step by step, he decreased their weight while increasing their reliability. In 1829 he built a kind of tractor to tow a carriage containing passengers. The boiler for this engine supplied steam at 120 pounds per square inch and it included a steam jet to increase the draught. When returning from Bath to London in August 1829 this vehicle covered eighty-four miles in just over nine hours, which included stops. A chassis and engine preserved in the Glasgow Museum of Transport are thought to be from one of Gurney's steam coaches.

Two years later Sir Charles Dance took over Sir Goldsworthy Gurney's coaches and opened a regular service between Gloucester and Cheltenham. Three or four times daily, coaches set out on the nine-mile journey which was covered in under an hour. Dance, however, encountered some strong opposition from owners of horse-drawn coaches and innkeepers, who even obstructed the road with piles of stones. The service had to be withdrawn after less than five months.

Several interesting steam carriages were produced in the 1820s; one of the first was built in 1822 or 1823 to the design of Julius Griffith of Middlesex. This had a two-cylinder engine driving the rear wheels, but the driving mechanism was unusual in that it included gears, so enabling the driver to change gear, as in a modern car. Another innovation was a condenser to cool the waste steam after it had been exhausted from the cylinders, thus converting it back into water which could be re-used in the boiler. This was a very interesting development, for one of the great disadvantages of the steam engine was its high consumption of water. A condenser employed in this manner conserved water, but it should not be confused with Watt's use of the condenser. In Watt's engines the condenser was an essential part of the engine, not merely a water-conserver. It was connected to the cylinder in such a way that the vacuum, formed by condensing steam in a sealed container, caused the piston to move. Later engines, and in particular marine engines, were fitted with condensers, not only to create a vacuum, but also to save water. Condensing steam in conjunction with steam under pressure was used to power the low-pressure cylinders of many compound engines, which therefore resembled the cylinder of a 'Cornish' pumping engine.

Despite Griffith's bright ideas for gears and a condenser, his

PLATE 29 *A Steam Fair. The inscription above the huge swing-boats reads 'Harry Lee's Steam Yachts, Columbia & Shamrock'.*

PLATE 30 *An American steam car of 1900, a Locomobile designed by the Stanley brothers. It has two inverted-vertical cylinders which can be seen—surrounded by lagging material—under the seat.*

PLATE 31 *A White steam car engine of 1909—a compound engine developing 15 horse-power. Part of the high-pressure cylinder has been cut away to show the steam valve arrangement.*

PLATE 32 *A Sentinel Steam Wagon, first used in the 1920's and now making appearances at shows and pageants.*

PLATE 33 *A 'Pacific' locomotive, the LNER* Silver Link *of 1935. Its speed of 112 miles per hour was later beaten by* Mallard *with 126 miles per hour.*

boiler was a complete failure, and even after modifications his steam carriage never really developed beyond the experimental stage. A similar fate befell David Gordon's carriage propelled by legs – for it was far too slow.

W. H. James of London and his Irish partner Sir James Anderson had some success and built several steam carriages including one, in 1825, with a four-wheel drive. Nevertheless, boiler troubles and heavy expenses brought their work to a halt. Another attempt to incorporate a four-wheel drive was made in the same year, by Timothy Burstall and John Hill who devised a mechanism which drove only the rear wheels in ordinary circumstances, but all four on hills. Burstall and Hill's steam carriage had another unusual feature; it was driven by two vertical cylinders fitted with 'grasshopper' beams, whereas most of the other designs made use of the neater direct-acting engine. Being very heavy, this machine could travel at a mere 3 or 4 miles per hour, and eventually it was discarded because of continual boiler failures.

All these boiler troubles emphasize the significance of Sir Goldsworthy Gurney's steam carriages, which actually worked. Even more successful were the designs of Walter Hancock, a London engineer who built nine practical and reasonably reliable vehicles between 1829 and 1836. Hancock started by patenting a boiler in 1827 which was a variation of the water-tube idea, called a 'cellular' boiler. In simple terms this was similar to a motor car radiator placed over a

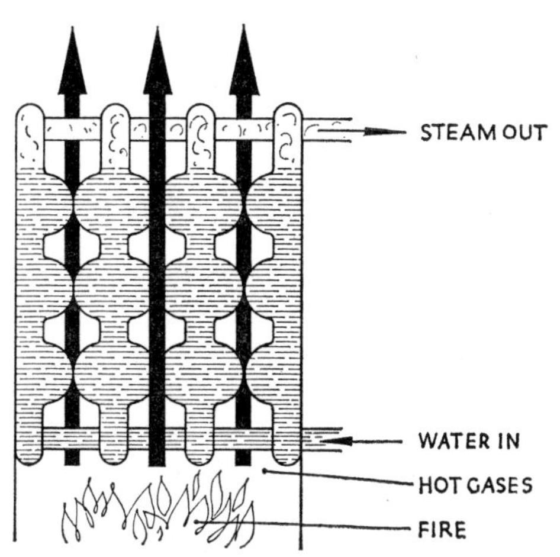

FIG 48 *Walter Hancock's cellular boiler installed in his successful steam coaches of the 1830's.*

fire; the water passing through the radiator would be heated up and converted into steam.

Hancock's first carriage was an experimental three-wheeler with a pair of oscillating cylinders driving the front wheel. It could carry four passengers and covered many miles in London despite the fact that Hancock was still carrying out experiments and improvements. His next steam carriage, called 'The Infant', was capable of carrying about ten passengers. This had a rear-mounted boiler and engine driving the rear wheels plus two steerable front wheels. In 1830 'The Infant' was considerably altered and enlarged to carry fourteen people. In February of the following year, this improved carriage was used to open a regular passenger service between Stratford in Essex and London – the first such service to be powered by steam. Incidentally, Hancock beat Sir Charles Dance by a few days for this honour.

Two of Hancock's most famous steam carriages were the *Enterprise* and *Autopsy*, built in 1833. The London and Paddington Steam Carriage Company ran the fourteen-seater *Enterprise* between Moorgate and Paddington, but Shillibeer's horse-omnibuses won the day and the company collapsed. Hancock himself renewed the service during the following year with *Autopsy* and *Era*. Between August and November some 4,000 passengers were carried at speeds of 12 miles per hour. An even more ambitious service was opened in 1836 covering Moorgate, Stratford, Paddington and Islington. Three carriages were employed; *Autopsy*, *Erin* and a large open charabanc called *Automaton* which could seat twenty-two passengers and travel at almost 20 miles per hour. During a period of twenty weeks the carriages covered a total of about 4,000 miles and carried nearly 13,000 passengers.

Most of Hancock's steam carriages had the same mechanical layout on which he produced three types of body – a traditional coach, an open charabanc and a closed omnibus. His boiler was at the rear of the vehicle with an engine-driven fan to assist combustion. The engine had two cylinders, mounted in a vertical position but inverted to drive a low crankshaft situated just forward of the rear axle. Chains transferred the drive from this crankshaft to the rear wheels. It was unusual at this stage in the development of the steam engine to adopt the inverted-vertical position for the cylinders, and the credit for popularizing this configuration is usually given to Nasmyth in 1850.

Hancock's vehicles were a great improvement on some of the first steam carriages, but they still presented problems. Breakdowns occurred fairly frequently, the ride was far from smooth, and they made a considerable noise. Their drivers required great skill to steer them on the bumpy roads and their boilers frequently needed fuel and water. A ten-mile journey would consume over a hundredweight of coke and ten gallons of water. Despite all these difficulties, one steam carriage made the journey from Stratford to Brighton in 8½ hours.

During the 1830s several successful steam carriages were built, some of which made long journeys. William Church's ornate three-wheeler ran between London and Birmingham in 1832, and Frank Hills' carriage of 1839 made journeys from London to Windsor, Brighton and Hastings. It even made a return journey to Hastings within a day in spite of stopping for water every eight miles.

By 1840 practically all the work on steam-powered road vehicles came to an end, not because of technical failures but for financial and social reasons. A steam carriage was not only an expensive item to build, but also its high fuel consumption had to be considered and, finally, road tolls were very heavy. Many of these tolls were biased against the steam carriage, which had to pay five or six times the amount paid by a horse-drawn coach. One of the most infamous examples was the Liverpool to Prescott road on which a coach was charged 4s., but a steam carriage had to pay £2. 8s. od. Naturally, the road owners claimed that heavy, high-speed vehicles would wreck their road surfaces.

The high operating costs might have been overcome if enough passengers could have been found, but the railways were progressing very rapidly and taking all the long-distance travellers, while new, comfortable horse-drawn cabs and carriages were appearing on the roads for the shorter journeys. The horse omnibus of 1829 has already been mentioned, after which came the famous two-wheeled Hansom Cab which was introduced in 1834, to be followed, a few years later, by the four-wheeled Brougham.

In addition to financial difficulties and rival forms of transport, the companies which ran steam-carriages had a further problem – many people were opposed to the steam engine in any form. Workers were afraid of any new machinery which might result in fewer jobs, and many members of the public objected to the noise, smell and smoke. The risk of being run over or killed by a bursting boiler also caused a certain amount of opposition. Sabotage, as encountered by Gurney, was fairly common on some roads.

A new interest in steam road vehicles grew in the 1860s and several promising designs emerged, but then came the Locomotive Act of 1865 – the notorious 'Red Flag Act'. This made it compulsory for a man with a red flag to walk in front of every powered road vehicle and limited their speed to 4 miles per hour in the country or 2 miles per hour in towns. As a result, the development of motor vehicles virtually stopped in Britain, but progress was made on the Continent.

The opposition to steam took some unusual forms. For example, although the steam engine had been used for pumping water since the days of Newcomen, there was considerable opposition to the first steam-powered fire engine. Basically a fire engine is a portable water pump, and the early examples were fitted with a pump worked by men – one even required forty-six men. Small engines were man-handled to the fire, but the larger ones were drawn by horses. In 1829 the first fire engine, with a steam engine to drive the pump, was built by the firm of Braithwaite and Ericsson. In the same year they also built the railway locomotive *Novelty* which was beaten by the *Rocket* in the Rainhill trials. The fire engine, designed by John Braithwaite, was demonstrated at several London fires. It had a vertical boiler, for which bellows provided a forced draught, and two horizontal cylinders driving the pumps.

Braithwaite's fire engine was drawn by horses, but in 1840 Paul Hodge produced a fire engine in which the steam engine powered not only the pump but also the road wheels. Both these designs were ahead of their times because steam pumps were not widely used until the 1860s, and self-propelled fire engines did not become a common sight until the end of the century.

At sea there was less violent opposition to the steam engine. For a considerable period there was no danger of the sailing ships being superseded because they were superior to their new rivals. Also, as both old and new forms of power could be incorporated into one vessel, there was no sudden change-over to steam as there had been on land.

Steamships probably made their greatest impact on the North Atlantic, which was to become the world's busiest long-distance route. The first crossing by a steamboat was made in 1819, when a little paddle-steamer with sails, called the *Savannah*, crossed from the United States to England. Although *Savannah* had a steam engine it was used only occasionally, and on the Atlantic crossing her sails provided the power for seven-eighths of the time.

The first crossing to be made under steam all the way caused great rivalry and excitement in 1838. Two companies were each preparing ships for the attempt and both were paddle-steamers, powered by side-lever steam engines plus sails. The smaller vessel, called *Sirius*, had been built for service across the Irish Sea, while her rival the *Great Western* was designed specifically for the North Atlantic by the famous engineer responsible for the Great Western Railway, Isambard Kingdom Brunel. A minor fire delayed *Great Western* and *Sirius* slipped out of harbour on 4 April to gain an advantage of several days. Despite headwinds, a mutinous crew and the fact that she ran aground at the entrance to New York harbour, *Sirius* tied up safely on 23 April. About four hours later the *Great Western* arrived. *Sirius* had won the glory, but she was too small for the North Atlantic, whereas *Great Western* continued in service between Bristol and New York until 1846.

Brunel's interest in steamships did not end with *Great Western*, for he went on to design two more world-famous vessels. The *Great Britain*, completed in 1843, had a double claim to fame – she was the first transatlantic liner to be powered by a screw propeller and the first to be built of iron. She had a remarkably varied career, but through all the changes and troubles Brunel's immensely strong iron hull remained intact. Even while being built, her design was changed from a paddle-steamer to incorporate the recently invented screw propeller. *Great Britain*'s engines consisted of four inclined cylinders driving an overhead crankshaft, a layout which required chains to transfer the drive down to the level of the propeller shaft. With the aid of her sails she could cruise at over 10 miles per hour carrying about 360 passengers.

After some demonstration voyages, the *Great Britain* made her first crossing from Liverpool to New York in the summer of 1845. The passage took under fifteen days and she continued on the North Atlantic route until September 1846, when she ran aground off the coast of Ireland where she remained for almost a year before being refloated. The iron hull withstood the ordeal, but some repairs were needed and her owners decided to sell her. The new owners refitted the *Great Britain* with oscillating engines, replaced the large funnel by two small ones and removed two of the six masts. After a further spell of service on the Atlantic she was transferred to the Australian route in 1852. By 1882 she was no longer competitive and had to be sold. Once again she was refitted, but this time her engines were removed and the *Great Britain* became a full-rigged sailing ship.

After a storm off Cape Horn in 1886 she limped into Port Stanley in the Falkland Islands and her days at sea were over. But this was not the end of the iron hull of the *Great Britain*, for it was used as a store for coal until October 1937. When it was no longer required for this purpose it was beached near Port William, where it remains today – a tribute to Brunel's skill in hull design.

Brunel had his failures as well as successes, but this is not surprising when the variety and advanced nature of some of his projects is considered. His steamship *Great Eastern* might be described as the grandest failure of the century. In 1852 Brunel suggested building a ship, not merely larger than any other, but more than five times the size of its largest rival. After many technical and financial difficulties the launching took place in January 1858 and the ship was named *Leviathan* – later to be changed to *Great Eastern*.

The *Great Eastern* was 680 feet long and designed to carry 4,000 passengers–although she never did. By comparison the *Queen Elizabeth II* is 963 feet long and has accommodation for just over 2,000 passengers. To propel such a vessel in the 1850s was a mammoth undertaking, and Brunel included three sources of power. Sails were carried on six masts, and two separate steam engines were installed to drive paddle-wheels and a screw propeller. The paddle-wheel engine had four inclined oscillating cylinders, each over 6 feet in diameter and with a 14-foot stroke. The propeller was powered by another four-cylinder engine located in the stern of the ship. To keep the crankshaft low, these cylinders were in a horizontal position and mounted across the hull. Because this layout was restricted by the width of the hull, the stroke of the direct-acting engine was limited to 4 feet, but the diameter of each cylinder was 8 feet. Naturally engines of this size required large quantities of steam and so the *Great Eastern* had ten boilers, five funnels and bunkers for 12,000 tons of coal.

The *Great Eastern* left the Thames for her maiden voyage on 7 September 1859, but Brunel had a stroke two days previously, and died just over a week later, aged fifty-three years. The strain of building a ship half a century ahead of its time had been too great.

There is no doubt that the *Great Eastern* was impressive and attracted sightseers wherever she went, but she could not pay her way. Fortunes had been lost building her and more were lost trying to operate her as a passenger liner. Eventually she was converted to lay under-water telegraph cables and from 1865 to 1874 she served a useful purpose. After a long period doing no work at all the *Great*

Eastern was sold for scrap in 1888 – but even scrapping her took three years.

Another of Brunel's failures, which was years ahead of its time, was the South Devon 'atmospheric railway' of 1847 to 1848. The idea was not new and was fundamentally sound. It was based on the principle of the atmospheric engine in which air (or atmospheric) pressure moved a piston sliding in a cylinder when a vacuum was created on one side of the piston. By connecting a train to a piston in a very long cylinder the train could be moved. Obviously, if the cylinder becomes a pipe several miles long, a normal piston rod cannot be used and, to overcome this difficulty, a connection between piston and train was taken through a slot in the pipe wall. This slot then had to be sealed to prevent leakage yet still allow the connecting link to pass along its length. A leather flap was tried, but this was to be the weak link in the whole system. In order to provide a vacuum, stationary steam engines drove pumps which extracted air from the 15-inch pipe on one side of the piston. These pumping stations were built every three miles along the twenty-mile route between Exeter and Newton Abbot.

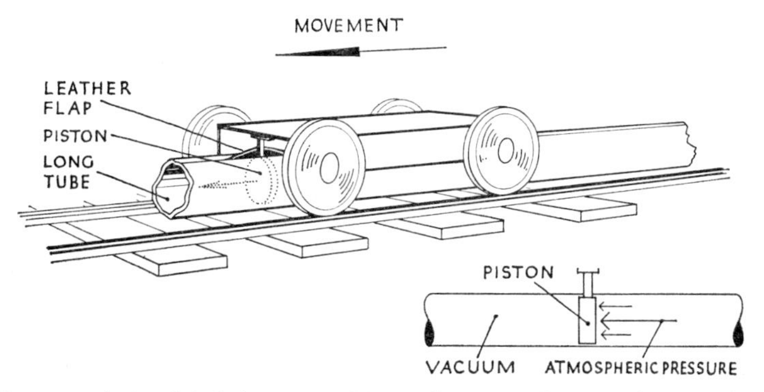

FIG 49 *A simplified diagram of Brunel's atmospheric railway built in South Devon, 1847.*

The line was opened in November 1847 and trains attained speeds of almost seventy miles per hour. The coming of winter caused havoc with the leather sealing flap, particularly when it became frozen, and other problems emerged, such as condensation inside the long slotted pipe. By the spring £25,000 was needed to repair the damage, and reluctantly Brunel recommended that the atmospheric railway should close.

Almost one hundred years later the slotted cylinder idea was revived to move aircraft instead of trains. It was incorporated into a catapult capable of launching naval aircraft from the deck of an aircraft carrier. In this modern version, high-pressure steam replaced the vacuum plus atmospheric pressure – as it did in the conventional steam engine. Another aspect of the atmospheric railway to be revived many years later was the idea of keeping the main source of power stationary, instead of in a locomotive. Fast, clean and quiet trains can result from this system, as was proved by Brunel and later confirmed by electric railways.

Although electric railways are often thought of as a twentieth-century innovation, they first appeared in the mid-nineteenth century. These early examples were little more than battery-powered prototypes, but they did demonstrate some of the advantages which could be obtained, particularly cleanliness and silence. Probably the first attempt to drive a vehicle by the power from an electric battery was made by two Dutchmen in 1835, only fourteen years after Michael Faraday's laboratory experiment which demonstrated the principle of the electric motor. During the late 1830s and 1840s several electric locomotives were built, and they achieved varying degrees of success. One of the most promising, built by an American called Thomas Davenport, was demonstrated on the Glasgow to Edinburgh line in 1842.

Davenport's locomotive was reputed to have hauled a load of 6 tons at a speed of 4 miles per hour, in which case it was comparable with Trevithick's first steam locomotive of 1804. Twenty-five years after Trevithick's prototype, practical steam locomotives were being built, but in the case of electric locomotives it took almost twice as long before they were widely used. Among the several reasons for this delay was the lack of experience with stationary electric motors before the first locomotive was built. Perhaps of more importance was the fact that the steam locomotive had become so well established that, if a rival was to succeed, it had to be considerably superior.

Steam engines to power factories, mines and transport have already been described, but in the 1840s a new use was found for steam power: it was harnessed to work on the land. A start had been made some time earlier, in 1811, when Richard Trevithick adapted his high-pressure engine to drive a threshing-machine. He even suggested that the engine should be mounted on wheels and moved about to power other threshing-machines, but this idea did not gain any real attention until the 1840s.

In 1840 the Royal Agricultural Society came into being and organized Royal Shows at various agricultural centres. These shows played a notable part in the history of steam power on the farm, because they gave the manufacturers an opportunity to demonstrate and advertise their new equipment. At the 1841 Royal Show, held in Liverpool, Messrs. Ransomes and May of Ipswich exhibited a portable steam engine – but like the early fire engines it had to be drawn by horses. Most portable steam engines to be used on a farm turned a flywheel which had a wide flat rim. Power was transmitted by wrapping a long leather belt around this flywheel and around a pulley on the machinery to be driven. This simple and effective method remained in use throughout the lifetime of agricultural steam engines – a period of almost a century.

In the following year Messrs. Tuxford and Sons of Boston, Lincs, produced a small steam threshing-machine with a boiler, engine and threshing-mill on the same wheeled-frame, but it was not a great success. Another threshing-machine was built in 1842 and demonstrated at the Bristol Royal Show. This was a development of Ransomes' portable engine shown the year before, but, in addition to driving a threshing-machine, this steam engine could also drive its wheels. A self-propelled steam traction engine was beginning to emerge. An unusual feature of this particular machine was the fact that it utilized a horse harnessed between shafts – purely to steer the front wheels. It could travel at speeds of up to 6 miles per hour and was said to be very manoeuvrable. Although this primitive traction engine was awarded a prize at the Royal Show, most farmers were apprehensive about this new form of power, fearing that sparks might set fire to their straw and buildings. Ransomes had to wait several years before steam engines were accepted by the farmers.

Most of the early portable engines had vertical boilers; then in 1845 the Lincoln company of Clayton and Shuttleworth built the first series of engines with a locomotive-type boiler. Both boiler and cylinders were in a horizontal position, but, unlike the locomotive layout, the two cylinders were mounted above the boiler. It was now only a short step to the first really practical, self-propelled traction engine. One claimant to this title was 'Willis' Farmer's Steam Engine' of 1849. This small vehicle was designed by Robert Willis, and built in Leeds at the Railway Foundry of E. B. Wilson & Co., in conjunction with Ransomes and May, who exhibited it at the Leeds Royal Show. The 'Farmer's Engine' utilized the railway locomotive layout of two cylinders below the boiler and between the front wheels. It could be used for towing purposes or as a stationary engine to drive a threshing-machine. For some strange reason no flywheel was fitted, with the result that when a belt-drive was required the rear end had to be jacked-up and the belt wrapped around one of the rear wheels.

From this point the traction engine developed its traditional layout with a horizontal boiler, long chimney and large rear wheels. Sometimes the cylinders were mounted below the boiler (undertypes), but the more common arrangement was to place the cylinders above the boiler (overtypes). A number of variations in design appeared which were to increase the usefulness of the traction engine. Despite the severe restrictions on any form of mechanically propelled road vehicle, traction engines were used on the roads to haul heavy loads. The most colourful examples were the showmen's engines which hauled steam fairs around the country. A number of passenger-carrying coaches were also towed behind steam tractors.

On the farm, elongated traction engines appeared with a huge drum under their boilers. Although these were ploughing engines, they did not tow the plough behind them – as did a horse – because their wheels would sink into any soft ground. To overcome this difficulty, the traction engines remained on the edge of the field and pulled the plough across by winding a long wire rope on to their steam-powered drums. Frequently two engines were employed, one on each side of a field, and they pulled a plough to and fro as they gradually moved along. There were many variations of the steam-ploughing technique, but in telling the story of the steam engine and its rivals space is unfortunately not available to describe all the applications of the steam engine.

CHAPTER 10

Theories and Improvements

(*c.* 1860) – Rival sources of power emerge

The latter half of the nineteenth century was a period of great scientific and engineering achievements. Darwin published his famous work *The Origin of Species*, and man learned to fly – if only in gliders. Steel was produced in large quantities and high-strength metals were developed. The Forth Bridge and the Suez Canal were built, and a telegraph cable was laid across the Atlantic. Electric light, the telephone, radio – all were invented; but more relevant to this story were the new and improved sources of power.

Electric motors for railways were mentioned in the previous chapter and gradually the use of electricity increased, not only to power motors but also as a source of heat and light. An electric motor, which is a very convenient and versatile power unit, differs from a steam engine because, in most cases, it is not a 'prime mover' – a term used to describe a mechanical device which converts natural energy into power. For example, a windmill draws its power from the wind, and a waterwheel is turned by the movement of water. A steam engine is a prime mover because it converts the heat energy in fuel into useful work. An electric motor poses a problem because it requires a supply of electricity which might be obtained from two very different sources. If it is produced in a battery by electrochemical action then the motor is a prime mover, but in most cases the electricity is produced by some form of dynamo or generator which, in turn, is driven by a prime mover such as a steam engine: in this case the electric motor is not a prime mover.

As a source of power, the electric motor became a serious threat to the steam engine, particularly for smaller applications such as machine tools. On the other hand, the fact that the electric motor was not a prime mover gave the larger steam engines a chance to extend their life by powering dynamos and generators. This was no easy task because electrical generators required high speeds, and in this field both steam and water turbines were increasing in popularity. Water turbines developed from waterwheels and, in regions where coal was scarce but water plentiful, they were installed in many factories and power stations. The first hydro-electric power stations were built in the mountains of Switzerland and the United States.

Water turbines can be divided into 'impulse' or 'reaction' types, the choice depending on the supply of water. A small reservoir high in the hills could supply a low-level power station with a limited quantity of water, but the great 'head' of this water resulted in it storing a large amount of pressure or 'potential' energy. This was converted into 'kinetic' energy in the form of a high-speed jet of water which was aimed at a wheel, not unlike a waterwheel. However, instead of flat blades, the turbine had spoon-shaped buckets which deflected the water back in the direction of the nozzle and thereby absorbed a maximum amount of the water's energy. Because this type of turbine utilized the impact of the jet on the bucket it became known as an impulse turbine. The most successful variant of this type was the Pelton wheel developed by an American engineer, L. A. Pelton, in the 1870s.

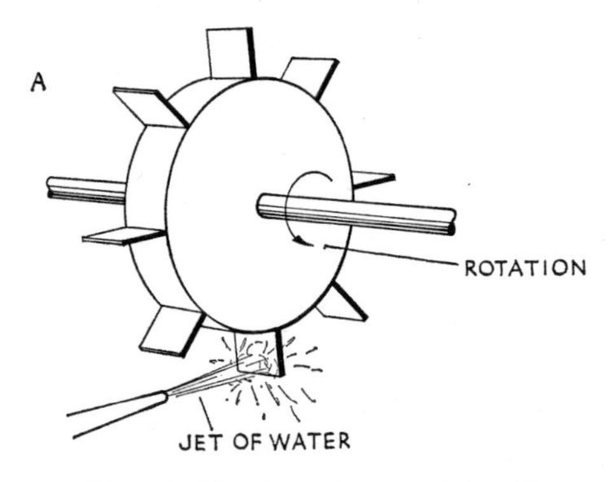

FIG 50 *Water turbine.* A *The impulse principle. If twin buckets are fitted in place of the flat blades, then this becomes a* Pelton *wheel.*

The other type of water turbine made use of the reaction principle described in Chapter 1 and demonstrated by Hero of Alexandria. There were several variations of the reaction turbine, but one of the simplest resembled an ordinary portable hair-drier. Instead of an electric motor forcing air through the drier the principle was reversed, and the action of water flowing through the turbine drove an electrical generator or some other machinery. One of the first practical water turbines to use this principle was built by a Frenchman, Benoit Fourneyron, who won a £250 prize for his design in

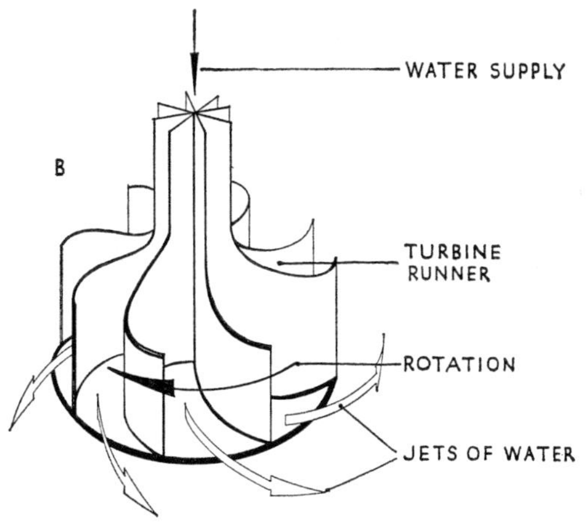

FIG 50 *Water turbine.* B *The reaction principle which utilizes the reaction of a jet.*

1827. During the 1840s an American engineer, James B. Francis, developed a new reaction turbine with an inward flow of water. (In terms of the hair-drier, the water went in where the air comes out.) Professor James J. Thomson of Belfast also developed an inward-flow turbine, which he patented in 1850 and called a 'Vortex' turbine. A small Vortex turbine was recently being used to power a woodworking machine in Northumberland. When this turbine was built in 1880 it powered the first hydro-electric generator in the United Kingdom.

Reaction turbines have a great advantage over their impulse rivals in that they do not require a great 'head' of water: one Vortex turbine used a head of only 3 feet. To compensate for this low head, large quantities of water were required, but at places such as Niagara Falls this was no problem. The Niagara hydro-electric power station began operations in 1895, despite opposition on the grounds that it would spoil the beauty of the Falls. Hydro-electric schemes frequently have to face opposition on this point because the most suitable sites for reservoirs are often in remote areas of great natural charm. Power stations driven by steam could be sited anywhere, but industrial regions were usually the most convenient.

Not only did water turbines affect the steam engine by providing competition; they also showed the advantage of the turbine principle which was later adopted for steam. A turbine, in which there are no reciprocating parts, runs very smoothly without vibration, making very high speeds possible.

A further rival of the steam engine during the nineteenth century was another reciprocating heat engine called a 'hot-air' engine. In principle this was similar to an expansive steam engine, but the piston was moved by expanding hot air instead of steam. Alternatively, if air is trapped in a closed cylinder fitted with a sliding piston and then heated, the air will expand and move the piston.

Sir George Cayley, who has already been mentioned for his aeronautical research, is generally accepted as the inventor of the hot-air engine. He certainly built several prototypes, the first being designed as early as 1807. Cayley's engine was basically very simple; hot air under pressure passed from a furnace into the cylinder where it expanded, thereby moving the piston. At the end of the stroke the air was allowed to escape. The chief difficulty with Cayley's engine was making mechanical parts which could withstand the high temperature needed to ensure a reasonable efficiency – a limitation which also restricted the success of later hot-air engines.

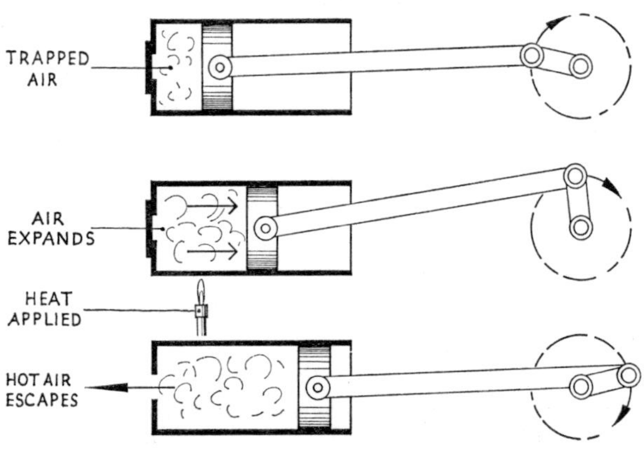

FIG 51 *The simplest form of hot-air engine.*

Scientists of the early nineteenth century still believed that heat was an invisible fluid called 'caloric', which flowed in and out of substances thereby changing their temperature. Because of this belief, the hot-air engine was also known as a 'caloric' engine. Following Cayley's lead, the Rev. Dr. Robert Stirling patented a hot-air engine in 1816. He made a significant improvement by using the same air, time after time, rather than wasting the warm air at the end of each stroke as Cayley had done. The air in Stirling's engine passed through a sequence of events called a 'closed-cycle', whereas

Cayley's engine worked on an 'open-cycle'. Expanding air provided the power stroke in both engines but, at the end of the stroke, the air in Stirling's engine passed through a 'regenerator' which absorbed most of the air's residual heat. This air was then compressed before entering the final stage which consisted of reheating. Heat was absorbed from two sources – the regenerator and an outside furnace.

Stirling and his brother designed a number of hot-air engines between 1816 and 1850, one of the largest being a 45 horse-power engine for a Dundee foundry. As the cylinder of this engine had to be heated to almost red heat, it did not have a very long life and it is said to have required three cylinder covers in four years – an example of a sound invention being held back because a suitable material was not available.

Some of the most spectacular hot-air engines were designed by John Ericsson, a remarkably versatile engineer and inventor, born in Sweden. After serving in the army, he moved to England in 1826 and patented a hot-air engine of the open-cycle type, pioneered by Cayley. In 1829, with John Braithwaite, he entered the locomotive *Novelty* for the Liverpool and Manchester railway trials, only to be defeated by Stephenson's *Rocket*. Ericsson patented a screw propeller for ships in 1836, but the British Admiralty was not interested. However, the Americans showed interest and, in 1839, Ericsson moved to the United States. Three years later the U.S.S. *Princeton* was built to his designs – the first screw-propelled steam warship.

Ericsson had built several experimental hot-air engines by 1853, when he designed four large engines for the paddle-driven ship *Ericsson*. Each cylinder was 14 feet in diameter with a stroke of 6 feet. The engines were designed to develop 600 horse-power, but unfortunately they were a dreadful failure and had to be replaced by steam engines. Ericsson was not deterred and, although he abandoned the idea of large hot-air engines, he continued to develop small ones. His ½ horse-power engine was so popular that he is reputed to have sold 3,000 of them by 1860.

From this time onwards it was realized that large hot-air engines were not a practical proposition, and they rarely exceeded one horse-power. The restriction on the maximum working temperature was always a handicap and the resulting engines were very heavy for their power. But they were safe and reliable – qualities which made them very popular for light work in a variety of places, from the home to isolated lighthouses.

Later manufacturers were divided; some followed Cayley's open-cycle principle while others developed Stirling's closed-cycle or regenerative engine. During the last quarter of the nineteenth century a number of very successful small hot-air engines were produced by Robinson in Britain, Heinrici and Lehmann on the Continent and Rider in the U.S.A. But when the compact and powerful internal combustion engine became established, the hot-air engine almost disappeared – although attempts to re-introduce it have been made from time to time. Even during the mid 1960s the Philips Company of Holland was developing a new hot-air engine.

Ericsson and his contemporaries, Brunel, Nasmyth, Perkins and others, were versatile engineers who progressed by a mixture of intuitive genius and hard work involving trial and error. The theoretical work often lagged behind practical achievements; as already mentioned, the false 'caloric' theory was still followed in the second quarter of the nineteenth century. A change took place in about 1850 when the theory of heat engines or 'thermodynamics' attracted attention from several eminent scientists.

One of the first major contributions to this relatively new science was made by a French soldier and physicist, Sadi Carnot. In 1824 he published a paper called *Reflections on the Motive Power of Heat*, but this scientific classic remained almost unnoticed for a quarter of a century. He stated the problem very clearly:

> 'Notwithstanding the work on all kinds of steam engines, their theory is very little understood, and attempts to improve them are still directed almost by chance.'

Carnot noted that heat engines could be simplified to three operations; the input of heat, the conversion of heat to work, and the rejection of unused heat. This in itself was not new, but Carnot went on to calculate a detailed cycle of operations which would give a perfect heat engine. Although the ideal 'Carnot cycle' could not be achieved in an actual engine, it provided a useful guide with which working engines could be compared.

Carnot, taking his theory a stage further, stated that a perfect engine must be 'reversible', and by this he referred to the cycle of operations. For example, heat flows into a normal engine and is converted into work, but if it were reversed and work put in, then heat would be pumped out and it would be a refrigerator. Carnot wrote his historic paper when only twenty-three years of age, and the fact that it was so near the truth is even more remarkable when it is remembered that he was using the false caloric theory of heat.

Perhaps his work passed unnoticed because of his youth and his early death due to cholera at the age of thirty-six.

In the 1840s an English scientist called James Prescott Joule carried out a series of experiments converting mechanical work into heat. This could not be explained by the caloric theory, and scientists were reluctant to accept Joule's results – the Royal Society even rejected a report on his work written in 1845. Two years later he gave a lecture to the British Association and afterwards wrote that '. . . the communication would have passed without comment if a young man had not risen in the section, and by his intelligent observations created a lively interest in the new theory. The young man was William Thomson.' Thomson was a brilliant scientist who had just been made a Professor at Glasgow University, despite the fact that he was only twenty-two years of age. In 1866 he was honoured for his work on the Atlantic telegraph cable – laid from Brunel's *Great Eastern* – and from then onwards he was known as Lord Kelvin.

A short time before Thomson met Joule he had 'discovered' Carnot's paper and was very impressed with its theory and logic. He was prepared to criticize Joule's work but, upon hearing the evidence, Thomson realized that Joule had made an important discovery and the two men became great friends. Thomson still could not bring himself to reject the long-standing caloric theory, but doubts entered his mind and he considered the problem.

While Thomson was pondering upon the nature of heat, a German scientist called Rudolf Clausius published a paper in May 1850 called *On the Moving Force of Heat and the Laws of Heat which may be Deduced Therefrom*. In this he considered the work of Carnot, Joule and Thomson, and then drew his own revolutionary conclusions. Rejecting the old caloric theory, he laid the foundations for the new science of thermodynamics.

The following year Thomson came to similar conclusions independently and published a paper *On the Dynamical Theory of Heat*. With two powerful supporters, the dynamical theory of heat was rapidly accepted in the scientific world, and this led to a new understanding of heat engines. The first textbook based on the new theories was called *A Manual of the Steam Engine and other Prime Movers*. It was published in 1859, and its author was another Professor at Glasgow University, William John Macquorn Rankine. Some ten years earlier Rankine had studied the nature of heat and published his theory, but this was overshadowed by Thomson's work.

Clausius, Thomson, Rankine and others developed the new subject of thermodynamics beyond existing engines. They were able to predict the behaviour of gases, and in particular steam, at high temperatures and pressures. The Laws of Thermodynamics were established and a peculiar property called 'entropy' was introduced to assist calculations. Because we cannot feel entropy, as we can temperature, it is difficult to understand, and even a simple dictionary definition is rather confusing. 'Entropy – Measure of the unavailability of a system's thermal energy for conversion into mechanical work.' However, engineers found they could use diagrams of temperature and entropy to assist their calculations, without a detailed knowledge of the scientific philosophy which had produced this strange property.

The birth of thermodynamics certainly helped to produce steam engines of advanced design during the second half of the nineteenth century, but there were other improvements based on new mechanical features. Of these perhaps the most significant were the improved valves which controlled the flow of steam into and out of the cylinder. Until this time many engines had slide valves which can be compared with a sliding door; then in 1849 a cylindrical rocking valve was patented which might be compared with a revolving door. Its inventor, George Henry Corliss, was not a trained engineer, yet despite this handicap he became a leading engine designer in America.

'Corliss' steam inlet valves were operated by a trip mechanism controlled by the governor and closed by the action of powerful springs. This arrangement controlled the speed of the engine very accurately – a factor which made Corliss engines very suitable for driving the machinery in textile mills. Many single-cylinder horizontal engines were built as mill engines, and Corliss valves were also fitted to compound engines and even to beam engines. As late as 1876, a huge beam engine was built by Corliss to drive machinery at an Exhibition in Philadelphia. It had two 40-inch-diameter cylinders and a flywheel 30 feet in diameter. Corliss engines were very economical and later versions developed several thousand horsepower. Their popularity continued well into the 1900s.

A very simple mechanical improvement was patented in 1852 by the locomotive engineer John Ramsbottom. This was the simple piston ring fitted around the piston to prevent high-pressure steam leaking between the piston and the cylinder wall. A number of sealing methods had been tried, ranging from rope to complicated spring-loaded devices, but Ramsbottom's split ring was delightfully simple and has still many applications, including the motor car engine. It consists of a metal ring cut at one place and sprung outwards. The ring sits in a groove cut around the piston and its own springiness holds its outer surface against the cylinder wall. Naturally, a small gap occurs at the cut in the ring, but leakage is reduced by fitting several rings.

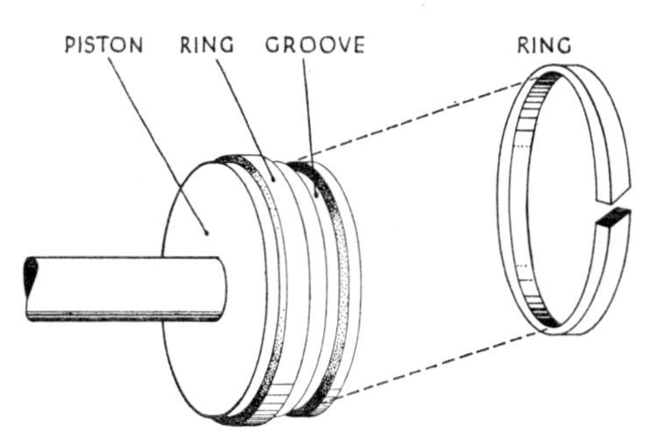

FIG 52 *A piston and piston ring.*

From a study of thermodynamics it became clear that engines must work at higher temperatures and pressures in order to increase their efficiency. One way of using this store of heat energy to the best advantage was to pass the steam through a small high-pressure cylinder, then into a larger, low-pressure cylinder. The reason for the low-pressure cylinder being larger was that the steam had expanded in the high-pressure cylinder and thereby increased in volume. Compound engines of this type were not new, for several compound beam engines had been built, but many of the early examples gave increased power without a marked improvement in the efficiency. One measure of efficiency is the amount of coal burnt for a particular engine power. This factor is vitally important at sea, where fuel has to be carried for a complete journey, and marine engine designers were often leading the way with new ideas.

The first sea-going vessel with a compound engine was built in 1854, but the first successful compound locomotive did not appear until over twenty years later, when Anatole Mallet introduced one in France. The ship in question was the screw steamer *Brandon*, and her compound engines were designed by John Elder of Randolph, Elder & Co. On her trials, the *Brandon* consumed only three-quarters of the usual amount of coal burnt by a ship with simple steam engines.

Most compound marine engines had vertical cylinders mounted in the inverted position over a fore-and-aft crankshaft, which drove the screw propeller directly. This was an ideal layout for a screw steamer and it remained popular for almost a century. The same layout was also used on land for stationary engines driving mills and power stations. As the high- and low-pressure cylinders were side by side and the steam passed across from one to the other, these engines were called 'inverted-vertical cross compound engines'.

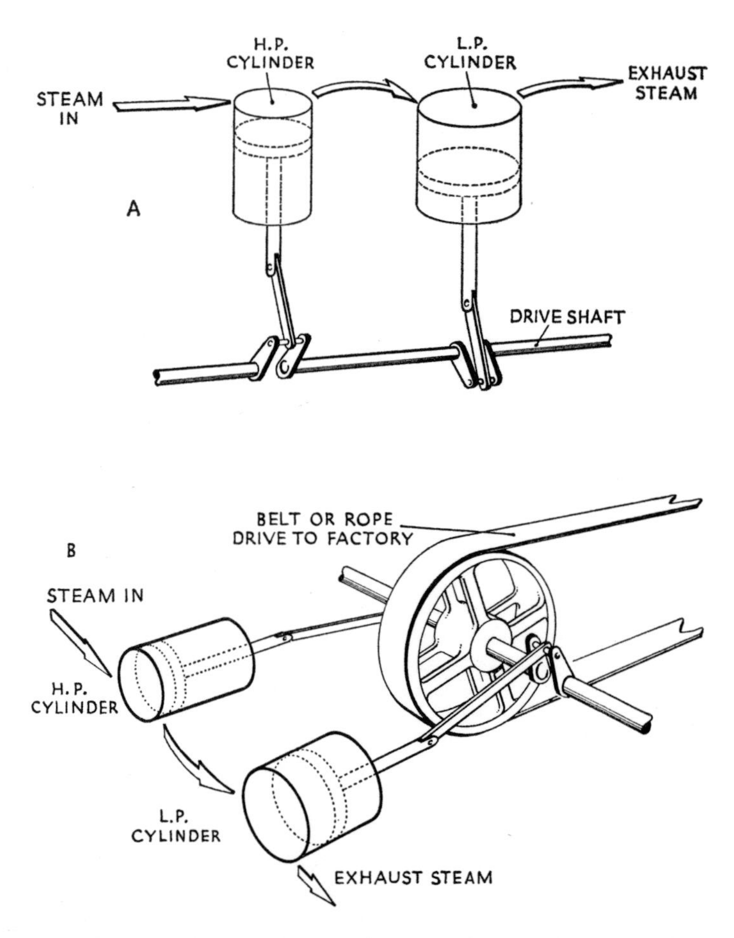

FIG 53 *Compound engine layouts.* A *Inverted vertical engine.* B *Horizontal cross compound engine.* C (*On opposite page*) *Horizontal tandem compound engine.*

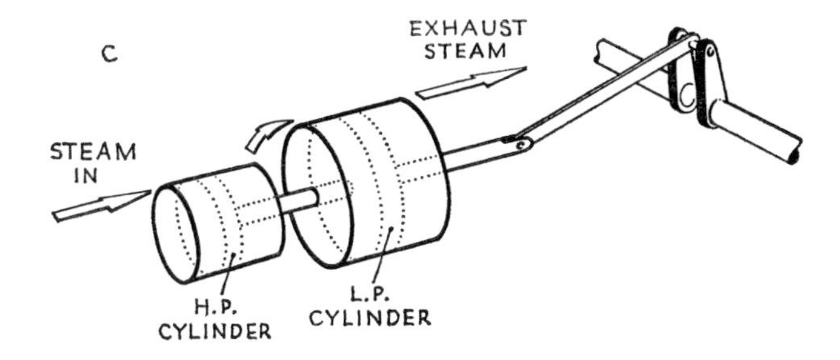

Horizontal cross compound engines were even more popular on land and were to be found in almost every industry. A large pulley was usually fitted between the two cranks, and from this belts or ropes drove the machinery. Engines ranged in size from a few horse-power to figures as high as 4,000, and there were many variations in details such as operating pressures, valves and condensers.

Another type of horizontal compound engine had its cylinders in line and so became known as a 'tandem' engine. The two pistons were linked by a rigid piston rod, and only one of the pistons needed to be connected to the connecting rod and crank. This layout was widely used for smaller engines, particularly where a narrow installation was required. Once again cotton mills were regular users.

An unusual compound engine was developed in America for use in power stations. This had a horizontal low-pressure cylinder and a vertical high-pressure cylinder or vice-versa. These very large engines were later introduced into steel works for driving the rolling mills.

To make full use of compound engines, steam at higher pressures and temperatures was required – a fact which led to the more widespread use of water-tube boilers and steam superheaters. Superheated steam was not new – Jacob Perkins had produced it in 1823. When superheated steam was first used it was primarily intended to remove any water and produce dry steam, because wet steam condensed very easily in a cylinder, with a resulting loss in efficiency.

During the 1860s several designs for water-tube boilers were produced, and one of the most spectacular was by Loftus Perkins, grandson of Jacob. The water passed through horizontal tubes arranged in a criss-cross pattern over a fire. A pressure of 2,000 pounds per square inch was recorded on test, in a boiler intended to operate at a figure of 500. However, as 200 was considered to be the maximum safe pressure at the time, only a few of these boilers were built.

A small, fast-steaming, vertical boiler, which incorporated water-tubes used in an unusual way, was patented by E. Field and Messrs. Merryweather and Sons in 1862. The upper half of an upright cylinder contained water, while a fire burned in the lower half. From the bottom of the water tank a large number of vertical tubes hung down, resembling fingers pointing at the fire. These were in fact two tubes, one inside the other, with the outer one sealed at the bottom. Water in the outer part of each 'finger' was heated and rose into the tank, while cooler water flowed down the inner tube to replace it. This boiler was used for some passenger vehicles and fire engine pumps.

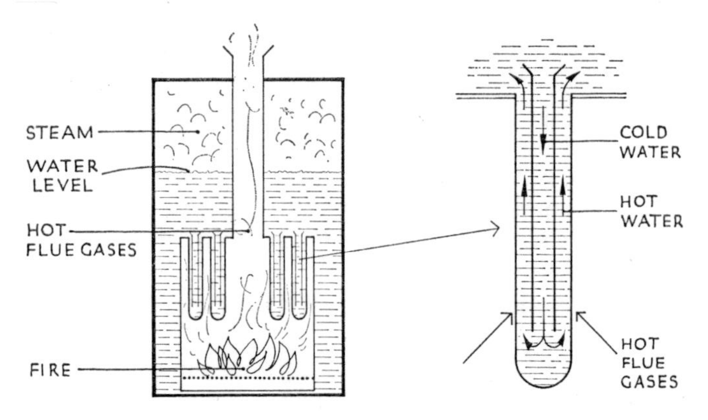

FIG 54 *A Field and Merryweather boiler—a small, fast-steaming boiler.*

Probably the most famous water-tube boiler of this period was the one patented by George Herman Babcock and Stephen Wilcox in the United States during 1867. This had a large number of parallel tubes inclined to the horizontal, with a fire below and a tank of water above. The Babcock and Wilcox boiler remained popular for many years and was easily adapted to take a steam superheater.

As boilers improved, the next advance was to make better use of the high pressures by expanding the steam in three stages using three separate cylinders one after the other; consequently this was called a triple-expansion engine. The idea was patented in France by Benjamin Normand in 1871, and a river-boat was fitted with the new engine. The first British triple-expansion engines were fitted into a ship, the S.S. *Propontis*, in 1874 and were designed by A. C. Kirk who worked for John Elder & Co. The installation was very similar to the earlier compound engines, but with three inverted vertical cylinders alongside each other.

Triple-expansion engines grew in popularity and eventually they became the standard form of power for ordinary merchant ships. Even during the Second World War these economical and reliable engines were fitted to the thousands of 'Liberty' ships built in the United States and supplied to Britain. On land, this type of engine powered pumps at some water-works, and it was used in mills, but

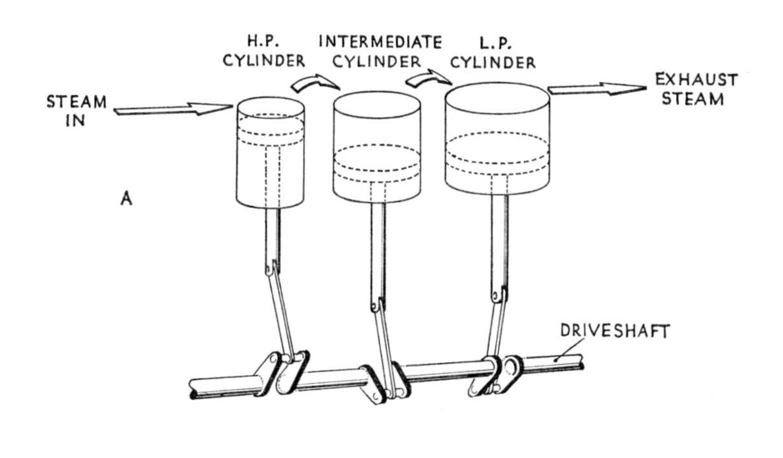

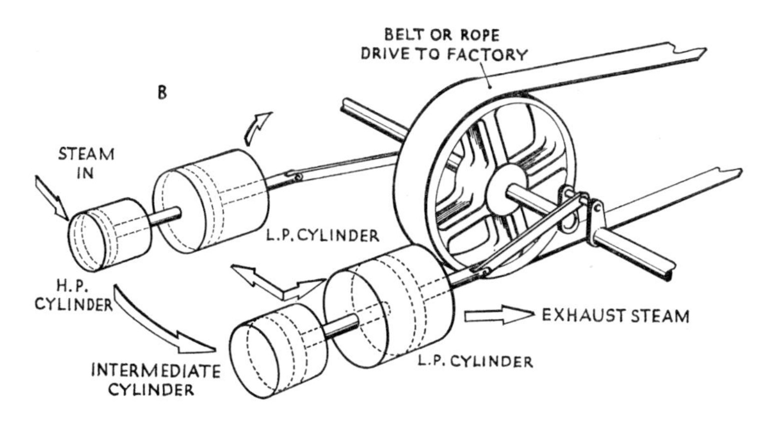

FIG 55 *Triple expansion engines.* A *Inverted vertical engine.* B *Horizontal, four-cylinder engine.*

the more popular triple-expansion engine for industrial installations was the horizontal version. Several designs were built, including one with four cylinders which was perhaps the most widely used. The cylinders were arranged in tandem pairs, side by side, with two low-pressure cylinders nearest the crankshaft. High-pressure steam was expanded in one of the far cylinders and then exhausted across to the other far cylinder for the intermediate stage. By this time, the steam was decreasing in pressure and increasing in volume to such an extent that it was divided, and half was fed into each of the two low-pressure cylinders. Although there were four cylinders, the steam was expanded in only three stages, but by the end of the century four stages were introduced in the quadruple-expansion engine.

By 1885 the steam engine was over one hundred and seventy years old and during all those years its efficiency had continued to improve. Some idea of this progress can be gauged by considering the quantity of coal burnt by a typical 50 horse-power engine in one hour:

Date	*Engine*	*coal consumed*
1712	Early atmospheric engine (Newcomen).	1600 lb.
1772	Improved atmospheric engine (Smeaton).	850 lb.
1776	Watt's engine with separate condenser.	450 lb.
1834	Cornish engine.	150 lb.
1870	Compound engine, horizontal.	100 lb.
1885	Triple-expansion engine, vertical.	75 lb.

From this table it can be seen that improvements in the early days produced dramatic reductions in the amount of coal used, but by the end of the nineteenth century it was becoming more difficult, and more expensive, to bring about a relatively small improvement. The steam engine was reaching the peak of its efficiency, yet it used only about one fifth of the heat energy stored in the fuel. This gave it a 'thermal efficiency' of 20 per cent which was not very good when compared with the relatively new internal-combustion engines. A gas engine was giving an efficiency of 25 per cent and, a few years later, a diesel had reached 32 per cent. These figures foreshadowed the end of the reciprocating steam engine.

CHAPTER 11

High-Speed Steam Engines

The challenge of the petrol engine

Conventional steam engines of the 1870s rarely rotated at speeds in excess of 100 revolutions per minute; therefore, if higher speeds were required, gears or pulleys and belts had to be employed. This was not an ideal arrangement because of practical difficulties, and not least the extra frictional resistance. Faster engines would be preferable, but there were several problems to be overcome. Balancing was a major difficulty because a large piston moving to and fro at high speeds could shake an engine, and even a building, to pieces. Then, too, there was the question of valves to control the flow of steam; as speeds rose the slide-valve and the more recent Corliss valve ceased to be effective. Finally, lubrication created a double problem as higher speeds caused more rapid wear and made oil distribution more difficult.

One of the first high-speed engines was patented in 1871 by Peter Brotherhood in England. The engine had three cylinders radiating in the form of a Y from a central crankshaft. (This 'radial' layout was later used with great success in aircraft petrol engines.) Each cylinder of Brotherhood's engine operated at the same steam pressure and was single-acting. Since the days of Watt, almost all steam engines had been double-acting, but in high-speed engines very severe stresses and strains were imposed by the piston pushing first one way, then the other. One way of reducing the severity of this loading was to keep it acting in one direction only, and this could be achieved by making the engine single-acting. By doing this, Brotherhood was able to attach the connecting rods to the pistons, thereby eliminating piston rods and crossheads. Not only was any saving in weight of the moving parts a great advantage in a high-speed engine, but also the radial layout gave a better-balanced and smoother-running engine than the conventional designs.

Brotherhood's engine was made in a range of sizes, all of which were relatively small with cylinders of up to 7 inches in diameter. The smallest engine developed 1¼ horse-power at 1,000 revolutions

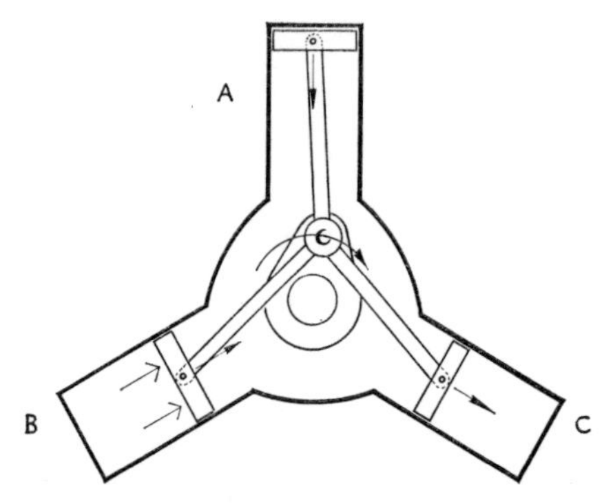

FIG 56 *Brotherhood's three-cylinder high-speed engine. Cylinder* A—*beginning working stroke. Cylinder* B—*nearing end of working stroke. Cylinder* C—*return stroke.*

per minute, while the largest produced 55 horse-power at 500 revolutions per minute. They were used to power high-speed machines, dynamos and even fans. The same layout was later adapted to be powered by compressed air and high-pressure hydraulic fluid in place of steam. An air-driven motor became the standard power unit for torpedoes, and it was only a short step from an air-motor to an air compressor.

During the 1880s activity in the world of engineering reached fever-pitch, and it was a decade of fierce competition. Probably the most exciting events were the railway races from London to Edinburgh in August 1888 between the trains operating on the rival East and West Coast routes. Each morning at ten o'clock the trains left London and raced to Edinburgh. Their latest times and speeds made headline news, but the difference between the two trains was very slight and their respective best speeds, including stops, were 52·7 and 52·4 miles per hour. Strangely enough, the locomotives which gave the East Coast route its slight advantage were old-fashioned 'single-wheelers' designed by Patrick Stirling. One of these engines, which did not actually race, is preserved in The Railway Museum, York, while at the Glasgow Transport Museum a locomotive which did race – on the West Coast route – can be seen. Although these were magnificent locomotives, the idea of a single pair of very large driving wheels was rapidly being superseded by several pairs of driving wheels coupled together.

The reciprocating steam engine, after a long period of virtually no opposition, was being challenged by water tubines, hot-air engines, gas engines and electric motors. Then in the 1880s petrol engines, oil engines (later to become 'diesels') and steam turbines were invented. In order to remain competitive as a source of power for electrical generators, the reciprocating steam engine had to increase its speed. Brotherhood's engine was a step in the right direction, but still larger engines were required.

A very successful high-speed engine was patented in 1884 by Peter William Willans who had carried out many engine trials and studied thermodynamics in order to explain the results. Like Brotherhood's, his engine was single-acting, but the cylinders were mounted in the inverted-vertical position. This made balancing more difficult, and to overcome the problem, Willans installed a special cylinder which acted as an air-cushion to soften the blow at the end of a stroke. A whole range of engines was produced including simple, compound and triple-expansion, and where two or more stages were required the cylinders were arranged in tandem one above the other, with the high-pressure cylinder at the top. If one tandem set did not provide enough power, then up to three sets could be built on to one crankshaft. A typical Willans engine developed 40 horse-power at 400 revolutions per minute, but engines of over 2,000 horse-power were eventually built.

Although Willans solved the balancing problem with the cushion of air, he was still faced with the question of a suitable valve mechanism and lubrication. Adequate splash lubrication was achieved with an oil bath in the crank-chamber, but Willans devised a very unusual valve arrangement. All the pistons in one tandem set were carried on a hollow piston rod, through which steam was supplied to the cylinders. Naturally the flow had to be controlled, and this was achieved by cutting holes in the rod which could be opened and closed by the action of a series of piston-valves sliding up and down inside the hollow rod. The piston-valves inside one rod were linked together and moved by an eccentric on the crankshaft. For obvious reasons this was sometimes called a central valve engine and it was widely used to generate electricity – until superseded by the steam turbine in the 1900s.

In 1890 a high-speed engine was built which soon rivalled Willans' design, although it appeared to be a conventional double-acting engine of the inverted-vertical type. It developed 20 horse-power at 625 revolutions per minute and its manufacturers were Belliss and Morcom of Birmingham. The secret of this contradiction of previous ideas lay in its lubrication system. Instead of relying upon oil splashing on to the bearings it had an oil pump which forced oil through tiny holes and into all the important bearings. This forced-lubrication system was the work of Albert Charles Pain, a designer at the Belliss works. Improvements followed, and eventually triple-expansion engines of 2,900 horse-power were built. Although the Belliss and Morcom engine, like the Willans, was superseded by the steam turbine in power stations, the idea of forced lubrication was incorporated into many other steam engines. These were sometimes called self-lubricating engines, and the idea later spread to turbines and internal combustion engines, so it was a very important development.

In the meantime boilers were being improved in step with the advances in steam engine design, and in 1889 the Babcock and Wilcox boiler was challenged by another American water-tube design patented by A. Stirling. The new boiler had a complicated layout with three upper and two lower drums connected by a large number of near-vertical water tubes. These were heated by a fire and flue gases, the latter winding their way through a maze which extracted as much heat as possible. By the end of the century, boilers were being made with mechanical stokers, heaters to heat the incoming water, and steam superheaters.

The year 1884 was a notable one in the history of steam power because in that year the first practical examples of both a steam turbine and a high-speed internal-combustion engine were produced. Steam turbines are described in a later chapter, as they are part of the story of steam power. When the internal-combustion engine first appeared it was closely related to the steam engine in both its method of operation and appearance.

Early experiments by de Hautefeuille, Huygens and Papin have already been mentioned, but the first practical internal-combustion engine was not produced until about 1860, when a French engineer called Étienne Lenoir made a gas engine. This resembled a horizontal double-acting steam engine complete with a cylinder, piston, connecting rod and crank, but in place of high-pressure steam a mixture of gas and air was ignited by a spark. The burning gases expanded, moved the piston, and turned the crank. Any exhaust gases were expelled by the piston making its return stroke under the pressure of burning gases on the opposite side. A Lenoir gas engine of 1860 is displayed in the Science Museum, London.

Several improvements to increase the efficiency of the gas engine followed, including an 'atmospheric' version, but the most notable development was the introduction of a compression stroke to increase the pressure of the mixture of gas and air before it was ignited. Credit for the invention of this 'four-stroke cycle' is usually given to a German engineer, Dr. N. A. Otto of the firm Otto and Langen, although he was only one of several advocates of the idea. Gas engines using the 'Otto-cycle' were made from 1876 onwards and they gained popularity very rapidly in many parts of the world, replacing small steam engines.

Gas engines were a great success as a stationary source of power, but they were not very suitable for vehicles because they were slow and, naturally, they required a supply of gas. Petrol vaporizes very easily, and it became obvious that petrol vapour could replace gas. In 1881 an employee of the firm of Otto and Langen left to produce his own, high-speed petrol engine. This engineer, called Gottlieb Daimler, had his engine working by 1884, and a year later he fitted it into a primitive motorcycle. In the same year, yet another German engineer, called Karl Benz, built a motorized tricycle quite independently of Daimler's experiments.

In 1886 Daimler went a stage further and made a four-wheeled car with a high-speed petrol engine. This car, which in many ways was the forerunner of the modern motor car, is preserved in a museum at Munich.

In the late 1880s a French company bought the rights to use Daimler's engine and a few years later their factory produced its first car under the name of Panhard-Levassor. By 1891 they were building cars regularly and selling them to the public. This was the birth of the motor industry which was to grow to giant proportions in twenty years.

About the same time as the petrol engine was being introduced, another type of internal-combustion engine emerged as a practical power unit. This was the oil engine which used low-grade fuel oil and in its early stages closely resembled the petrol engine. One of the first oil engines to be a commercial success was produced by the Priestman Brothers of Hull in 1885. Fuel was sprayed into the cylinder under pressure, heated and ignited by an electric spark.

Later engines dispensed with the spark but retained some form of heating to ensure an explosion of the fuel and air. In 1892 Dr. Rudolf Diesel of Germany patented a compression-ignition engine which dispensed with a heating device. The air in the cylinder was compressed by the piston to a very high pressure, then fuel was injected into the cylinder, the high pressure causing the fuel to ignite spontaneously. Dr. Diesel suggested that coal-dust would be a suitable fuel, but his idea was developed as an oil engine by the German firm M.A.N. in 1895 and later by others. Even these early diesel engines had a thermal efficiency which was 50 per cent better than a steam engine. Their future was assured.

CHAPTER 12

Steam Engines in the Twentieth Century

The end of the line

In the early years of the twentieth century Queen Victoria died and the Edwardian age began. It was a time of change in many spheres – social, political, military, industrial, and of power in both senses. The new sources of mechanical power, which have already been described, took some time to oust the steam engine from its dominating position of the Victorian era. For many years the old and the new ran side by side.

In the early 1900s the faithful steam engine was very much in evidence, despite the predictions of the thermodynamicists and their figures of thermal efficiency. After all, the most economical engine in the world would be of little use if it broke down frequently, and by this time the steam engine was extremely reliable. Although the 1900s marked the beginning of the end for the steam-powered piston engine, the steam turbine was growing in popularity. This new form of power shared one important feature with the old: they both required a boiler to produce steam, and so boiler design continued to improve during the present century. Any increase in boiler efficiency benefited both reciprocating engines and turbines alike, but, whereas the turbine was also being improved rapidly, the engine had almost reached its limit.

One improved engine design to appear in the early twentieth century was the 'Uniflow' engine of 1908. The idea behind this design, which eliminated exhaust valves, was not new, and Jacob Perkins' experimental single-acting version of 1827 has already been mentioned. This early engine used steam at the astonishing pressure of 1,400 pounds per square inch. A further attempt to produce a 'Uniflow' engine was made, in 1881, by Leonard Jennett Todd, but his double-acting design suffered from constructional difficulties. The successful engine of 1908 was designed by Professor Stumpf of Charlottenberg, and although this engine resembled Todd's, it incorporated drop (or plug-type) valves in place of slide valves.

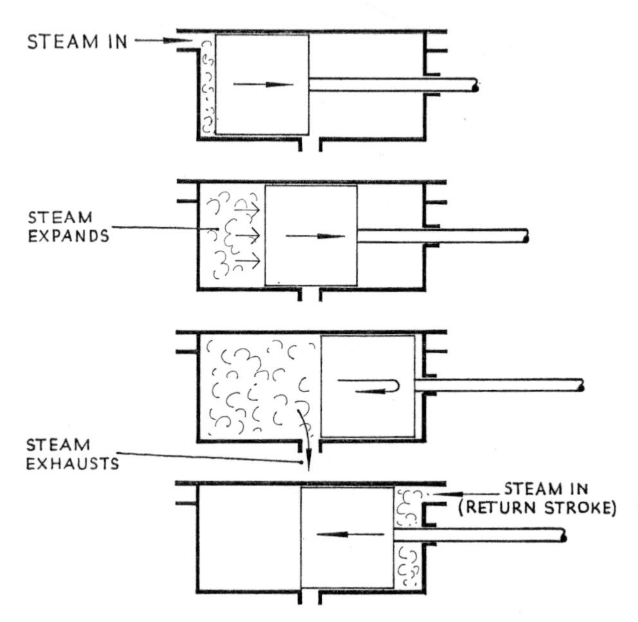

FIG 57 *The principle of a 'Uniflow' steam engine.*

Very high-pressure steam entered the cylinder of the 'Uniflow' engine through one of the two inlet valves, which were situated at either end, and moved the piston in the usual way. As the piston reached the end of its stroke, holes (or ports) in the cylinder wall were uncovered and the waste steam escaped. By making the piston very long, the same holes could be used to exhaust the steam at the end of the next stroke in the opposite direction. This arrangement had a great advantage because the ends of the cylinder could be kept very hot while the centre remained cooler, whereas in a conventional engine the temperature of one end alternated between the hot incoming steam and the cool exhaust steam. This one-way system led to the name 'Uniflow' and resulted in a reduction of the wasteful condensation which occurs when hot steam enters a cool cylinder.

Single-cylinder Uniflow engines were as efficient as triple-expansion engines, yet far simpler and smaller. They were built over a period of about twenty years and powered mills, electrical generators, colliery winding engines, ships and locomotives.

By 1900, the compound principle had been taken a stage further and quadruple-expansion engines were being built. The steam passed through a series of four cylinders; high-pressure, first inter-

PLATE 34 *A steam-powered swing bridge opened in 1885 to carry a railway line across the River Forth upstream from the famous Forth Bridge.*

PLATE 35 (*Top*) *An early Parsons' axial-flow turbine driving a dynamo. This unit was built in 1890 to provide electric power for H.M.S.* Gossamer.

PLATE 36 *The* Turbinia *at the Museum of Science & Engineering, Newcastle upon Tyne.* Turbinia *was the first vessel to be driven by a steam turbine, having been designed in 1894 by Sir Charles Parsons.*

PLATE 37 *A model of a 1930 Brush-Ljungström axial-flow turbine, cut away to show the blades mounted on contra-rotating discs. The right-hand disc is made of transparent material to reveal the actual blades.*

PLATE 38 *Reduction gears for the large steam turbines installed in the* Queen Mary.

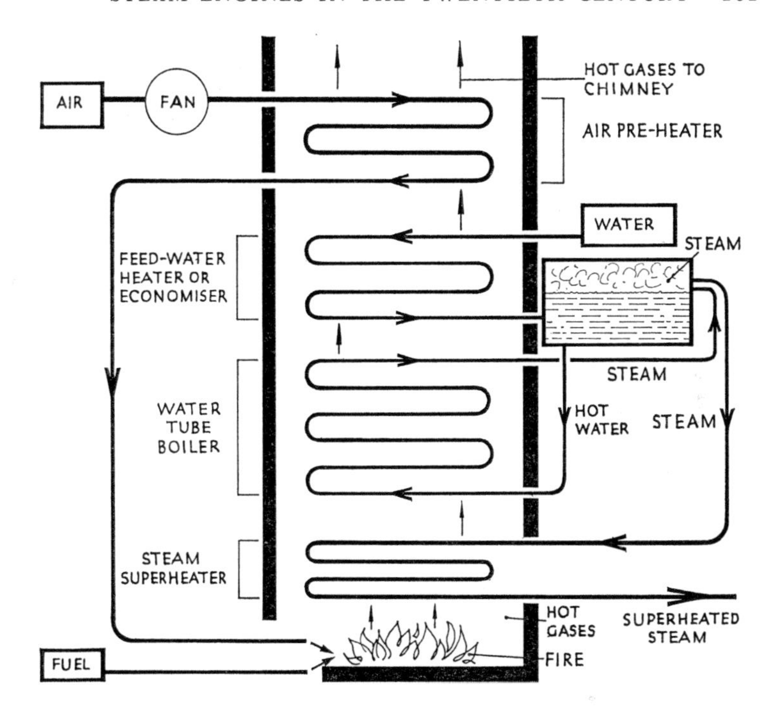

FIG 58 *A diagrammatic layout of a modern boiler showing the various systems.*

mediate, second intermediate and low-pressure. Several arrangements were possible, but for marine engines the four cylinders were usually arranged side by side in the inverted-vertical position. Because engines of this type were complicated and expensive, their application was not widespread – except as marine engines.

The Uniflow and the quadruple-expansion designs represent the last major developments in the history of the steam engine. Of course minor improvements were made and existing designs were built for many years, but little actual progress was made. By contrast, many changes in boiler design were introduced, and boiler efficiencies rose from about 70 per cent to over 90 per cent during the first half of the present century. This substantial increase in efficiency was not so much due to revolutionary new principles as to a careful study and gradual improvement of each aspect of a boiler's operation. For example, the 'Yorkshire' boiler of 1906 was developed from the 'Lancashire' type but with improved fire-tubes and flues.

The demand for high pressures and rapid steaming led to the adoption of the water-tube boiler which basically consists of a fire under a number of water-carrying tubes. The problem is to extract as much heat as possible from the fuel and transfer this to the steam. A boiler has to be supplied with three commodities – fuel, air and water, and from these it produces steam, plus hot flue gases with smoke and waste material such as ash. The heating system and the water-to-steam circuit of a modern boiler are more easily understood when considered separately.

Ordinary coal thrown on to a fire by a man with a shovel was the traditional method of firing a boiler, but by the 1900s some more advanced methods were being introduced. Power stations and marine installations were in the forefront of mechanization, while the railways retained the fireman and his shovel to the end of the steam age. Many of the improvements in boiler design were invented long before they were widely used, and the mechanical stoker was no exception. A Manchester blacksmith called John Stanley patented one in 1822, but until the end of the century they were rarely used. An alternative was to burn a fuel which was more easily supplied to the furnace – such as coal gas, oil or pulverized coal. Some oil-fired locomotives had been built in 1886 for the Great Eastern Railway, but the high cost of fuel oil made them uneconomic. Naval authorities were less worried about cost and they began to introduce oil-fired boilers for warships in the 1900s. Oil was less bulky than coal, and this was a very important consideration in a warship.

When fuel burns it requires a supply of air, and one refinement is to force more air through by using a fan, while another is to heat the air in a pre-heater, an idea introduced in about 1918. This device is a type of 'heat exchanger' placed in the path of the waste flue gases to absorb some of their heat. A domestic radiator is a simple heat exchanger which transfers heat from hot water to air without the two substances coming into direct contact.

The heat of the flames in the furnace is not the only source of heat, as the process of combustion produces hot gases, which are wasted in a domestic open fire. In most boilers, however, the flue gases make a very important contribution to the water- and steam-heating process. Any heat not absorbed from the hot gases by the boiler is wasted up the chimney.

The second part of a boiler's operation concerns the water and steam circuit. Before the water enters the boiler proper it usually passes through a feed-water heater or 'economizer', which is another heat exchanger placed in the path of the hot flue gases. Hot water from the economizer passes into the water-tubes and is converted into steam. Sometimes the circulation due to natural convection is sufficient to keep the water flowing through the tubes, but more steam can be produced by fitting a pump and using forced circulation. Steam is collected at the top of the boiler and passed into a superheater, which is often placed in the hottest part of the furnace. By the time it emerges the steam might be at a pressure of over 1,000 pounds per square inch and a temperature of 1,000°F.

The modern boiler, with all its pumps, heat exchangers and automatic fuel supply, is vastly different from Newcomen's 'haystack' boiler, and this difference is obvious even at a glance because of the sheer size of the complete installation. Just one of the many boilers in London's Battersea Power Station, made by Babcock and Wilcox, stands over 120 feet high and delivers almost half a million pounds of steam in one hour. The pressure of this steam is 1,400 pounds per square inch with a temperature of just under 1,000°F.

Large water-tube boilers with all their complicated accessories are only installed in power stations and very extensive factories; most owners of steam engines were content to make do with less efficient and less expensive boilers. In 1913 a survey of stationary and portable boilers in the United Kingdom revealed that the Lancashire boiler was, by far, the most popular design: this was followed by the vertical boiler, the locomotive-type boiler, the Cornish boiler and the water-tube boiler in that order. Of these, the vertical and locomotive-type were probably best known to the public because they were used for the smaller portable engines, whereas the others were housed in boiler-houses.

In the early part of this century, steam engines were a common sight and not just as a source of power for road vehicles or locomotives. Large steam cranes were used to lift heavy objects at the docks, and smaller, portable steam cranes were adapted for a variety of work. Sometimes cranes were mounted on locomotives with separate cylinders to drive the wheels and crane. But the more usual steam crane was not self-propelled and consisted of a vertical boiler supplying steam to a two-cylinder engine. The pile-driver was very similar to the crane in the way it lifted a heavy object, but the pile-driver released its weight over a pile (or post) and thereby drove it into the ground.

Little steam-driven 'donkey' engines were to be found in many places driving winches, machines or pumps. These were usually reliable single-cylinder engines of very simple design. Another type of pump to be seen pumping water out of flooded civil engineering works or mines was the 'Pulsometer' steam pump – derived from the 'Miner's Friend' pump which was patented by Captain Thomas Savery in 1698. The basic principle remained the same, but the new version, invented by an American, H. Hall, in the 1870s, had greatly improved valves which operated automatically. Water was lifted by the action of steam condensing, and forced upwards by steam pressure – an operation which did not require a piston because the steam acted directly on the water. Pulsometers were built well into the twentieth century and proved to be very useful in an emergency, because they could pump water containing mud and even small solid objects. They did have two disadvantages; as a supply of steam was required, this meant a boiler, and they were not economical. Of course, where a contractor was using a steam crane or a pile-driver on a particular site he might have plenty of spare steam available and therefore the disadvantages were immediately overcome.

Civil engineering contractors used steam power in a number of ways, but the most widely known – and that best loved by small boys – was the steamroller. The first steam-powered road-rollers were actually traction engines towing a huge roller; then in the 1860s the first real 'steamrollers' were built to the traditional layout, which lasted for a century. A locomotive-type boiler was normally used and a number of improvements were introduced over the years, including compound cylinders, different valves and oil-fired boilers. The first internal-combustion engined road-rollers were introduced in the early 1900s, but steamrollers were not to be superseded easily: rollers were necessarily slow and heavy, and both these features suited the steam engine. Steamrollers were still being built after the Second World War, in fact between 1945 and 1950 almost 500 were built in Britain. These were mostly for export to India and the Far East where timber was readily available as fuel, but where diesel oil was scarce and expensive. Probably the oldest surviving steamroller in Britain is an 1882 Aveling and Porter which worked continuously until 1960, including operations on the M1 Motorway during 1959. It has been restored by Aveling-Barford Ltd., and is preserved at their Grantham headquarters.

The problem of determining the first example of any particular invention has already been mentioned several times, yet it is often even more difficult to say which was the last. One reason for this is the fact that something new is news, but something on the wane often passes unnoticed. Another difficulty is keeping track of the last few; for instance, a steamroller could be working in a remote part of the country without anyone locally realizing that all the others had been retired. Enthusiasts help to record many of the examples of the steam age, but not all items can be covered.

Traction engines are in no danger of becoming extinct because private owners and societies have bought most of the surviving examples to restore and run for pleasure. In the early part of this century, the traction engine was a relatively common sight on the roads of Britain, hauling heavy loads. Threshing-machines or trailers carrying roundabouts for a fair were probably the most widely seen loads. From about 1890 special showmen's engines were built for fairground work, particularly by Burrells of Thetford, and these were highly ornate and colourful machines. Technically they were similar to an ordinary large traction engine, but later versions were fitted with a dynamo to provide power for the fair's electric lighting. The dynamo was normally mounted on the front of the boiler and driven by a belt from the flywheel.

A fair of the 1900s included steam-driven roundabouts and switch-backs, swinging steam yachts and, of course, a steam-powered organ. The most famous organ builder was probably Ludovic Gavioli, an Italian who lived in Paris. One of his large organs cost up to £5,000, which represented a considerable outlay at a time when a house cost a few hundred pounds only. Gavioli's organs had a steam engine to pump air through their organ pipes, but an American called Arthur Dennis made an organ in which high-pressure steam was used in the pipes – with deafening effect.

Traction engines and steamrollers were by no means the only examples of steam-powered vehicles on the roads during the first quarter of the twentieth century; there were cars, buses, lorries and trams – all driven by steam. During the latter part of the previous century, British designers of road vehicles had been restricted by various Acts of Parliament. Some progress was made in Britain, but American and Continental engineers were forging ahead. In France, the first major road race was held in 1894 from Paris to Rouen – a distance of about seventy-five miles. Despite fierce competition from petrol-engined cars, a De Dion-Bouton steam car won at 12 miles per hour. In 1896 the British law, requiring a man to walk in front of every car, was withdrawn, and the motor car was, at last, recognized as a means of transport and not merely a large toy.

Car design was still in a fluid state in 1900 and there was great rivalry between three sources of power – electricity, petrol and steam. This rivalry is reflected in the World land-speed records which show that the steam car was by no means inferior. Electric cars completely dominated the early attempts and took the record to 65·78 miles per hour in 1899. This record stood for three years and then it was broken by almost 10 miles per hour by Léon Serpollet in his steam car. A few months later a petrol-engined car exceeded Serpollet's speed by a small margin. By a strange coincidence the first car to exceed the magical 100 miles per hour did so in 1904 – the same year as it was achieved on the railways. The following year another petrol-engined car took the record to just over 109 miles per hour; then in 1906 came another shock – a Stanley steam car travelled at 127·66 miles per hour. This record represented such a dramatic increase that it remained unbeaten for four years. But this was the end of the steam car in the land-speed record table and the internal-combustion engine became supreme.

The steam car was still very popular on the roads; in fact, at the beginning of 1908 there were more 'steamers' in the United States than cars with petrol engines. A typical American advertisement for a 1903 White steam car read:

> 'The Car renowned the world over, for reliability and for flexibility of control. Steam the motive-power, but not a steamer in the ordinary sense.
>
> An entirely unique system of steam generation with automatic fuel regulation. High economy compound engine, with condenser; there is therefore no exhaust and the water is used over and over again.
>
> 100 miles on one filling of water and gasolene.'

The references to the steam generator and gasolene indicate an unusual boiler, and this was, in fact, a 'flash' boiler. Steam was made instantaneously by passing a small quantity of water along a tube heated to a very high temperature by a petrol burner. Jacob Perkins had made a flash boiler in about 1823, but the short life of his iron tubes and other difficulties led to the idea being discarded.

In 1887 a new flash boiler was patented in France by Léon Serpollet who developed the idea of using nickel-steel tubes heated until they were red-hot. Serpollet's flash boiler produced dry superheated steam, ideal for the engine, and as the steam was produced instantaneously this reduced the warming-up time before a journey could begin. Since there was no reserve of steam under pressure, as in a normal boiler, there was no danger of explosion, but on the other hand there was no reserve for extra power on hills. It required considerable skill to control the steam output of Serpollet's boiler, but it was widely used and, as already mentioned, Serpollet held the world land-speed record in 1902.

The White Company improved Serpollet's boiler to ensure that sufficient steam would be available at all times and give 'flexibility of control'. The White steam generator, or flash boiler, produced superheated steam at 600 pounds per square inch and was used until about 1912 when the company ceased producing steam cars.

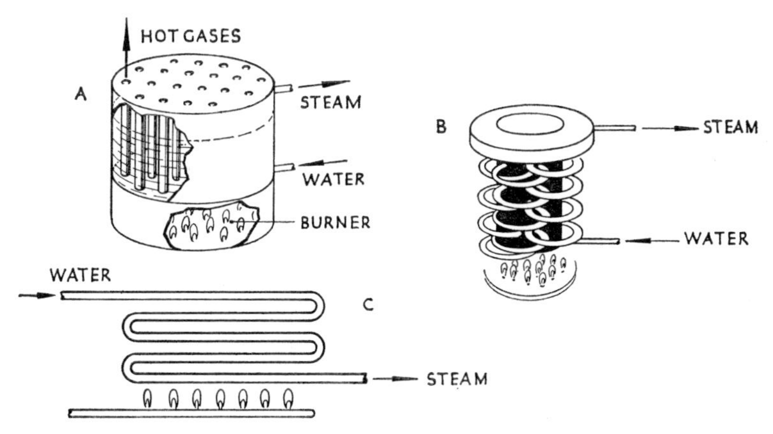

FIG 59 *Steam car boilers.* A *Stanley fire-tube.* B *Ofeldt water-tube.* C *White flash boiler.*

Fire-tube and water-tube boilers were also installed in road vehicles with considerable success, despite the long period they required to 'raise steam'. These boilers were fired by a variety of fuels including coal, coke, petrol and paraffin. Probably the simplest fire-tube boiler was the one fitted in the American Stanley steam car, which had a cylindrical water tank with vertical fire-tubes from top to bottom. A powerful petrol-burner was placed beneath this tank and the hot fumes passed up the vertical tubes.

Large water-tube boilers usually have straight or nearly straight tubes, but many automobile boilers had coiled tubes in order to economize in space. The Ofeldt boiler had a veritable maze of tubes. One exception to this rule was the boiler used in some of the later Doble steam cars in America. This had row upon row of vertical tubes through which water was pumped and it included a feed-water heater or 'economizer'.

Boilers were the vital part of steam car design because conventional designs, installed in factories, ships and locomotives, were far too large and heavy for light cars. A new range of boilers had emerged to suit the requirements of the motorist, but the actual steam engines were quite conventional, although different makers had individual preferences for certain cylinder layouts, valve gears and driving mechanisms. Serpollet's steam carriage of 1889 had twin cylinders in a horizontal position, while the 1903 White car had a two-cylinder compound engine mounted in an inverted-vertical position at the front of the car, driving the rear wheels through a propeller shaft. In 1917 the Stanley steam car had a two-cylinder horizontal engine, but each cylinder was simple and double-acting. The engine drove directly on to the rear axle and was controlled by D-slide valves in conjunction with the Stephenson link motion. The Doble engine of this period had a similar layout, but its cylinders were of the 'Uniflow' type. In outward appearance these steam cars were hardly distinguishable from their internal-combustion engined rivals – they even had a radiator at the front. This was installed to cool waste steam and condense it into re-usable water; it was therefore a condenser but one which did not affect the operation of the engine, differing in this respect from Watt's condenser.

The production of most steam cars ceased in the 1920s, although one or two – such as the Doble – survived until the 1930s. The reasons for this decline are complex, involving prejudice against high-pressure boilers, the complications of an engine plus a boiler, the low thermal efficiency of steam engines and, finally, the decision of American businessmen, including Henry Ford, to mass-produce cars with internal-combustion engines.

Commercial road vehicles powered by steam engines were probably more successful in Britain than elsewhere. If these had been developed from the British traction engine, then it would have been a logical story, but a large proportion were not: they were derived from the early lightweight steam cars which incorporated vertical boilers. Eventually heavy lorries were required, and by about 1905 designs based on the traction engine, with its locomotive-type boiler, were being built. But by this time many vehicles with vertical boilers were in daily use.

In 1892 steam delivery vans were introduced by some Paris stores and these vehicles, built by Le Blant, incorporated a Serpollet boiler. The first practical British steam van was built by John I. Thornycroft in 1896, and this was followed by a van made by the

Lancashire Steam Motor Carriage Company of Leyland – soon shortened to 'Leyland', a famous name still. This van burned oil fuel, but from 1899 Leyland vehicles burned coke as it was cheaper. Incidentally, as early as 1897 Thornycrofts built two steam dust-carts for use in Chiswick. Steam buses were tried in London between 1902 and 1909, but they were not a great success.

Steam lorries or wagons were firmly established in Britain by the end of the 1900s and there were over thirty companies engaged in their construction, compared with about six producing steam cars and more than seventy building petrol-engined cars. By the end of the 1920s severe road taxes on heavy vehicles and the increasing popularity of the diesel engine had reduced the number of companies building steam lorries to four. The names Foden, Garrett, Sentinel and Yorkshire had been synonymous with steam lorries since the early years of the century. In 1930 a Foden was built with pneumatic tyres, in place of the usual solid rubber. This typical example of the last generation of steam lorries had a vertical boiler behind the driver's cab and a horizontal engine under the chassis. It could travel at 40 miles per hour. After World War II most of the makers turned their attentions entirely to diesel engines, but a few steam lorries have been produced for countries where fuel-oil is scarce.

Steam-propelled trams and fire engines enjoyed a short period of success in the early part of this century, before steam was rendered obsolete by electricity and petrol respectively. One firm's name linked these very different vehicles – that of Merryweather and Sons of London. Their vertical boiler, which supplied steam for the pumps of horse-drawn fire engines, has already been mentioned. During the second half of the nineteenth century, Merryweather, and their rivals Shand, Mason and Co., were the leading manufacturers of horse-drawn fire engines.

The first steam-propelled fire engines were built in the 1890s, yet in 1903 even the steam-conscious firm of Merryweather and Sons built the first fire engine powered by a petrol engine. A few more steamers were built, but by 1918 their short life was over.

Steam trams began to make their mark in the 1870s when the only alternative for local journeys was a horse-drawn vehicle. Three types of steam tram were eventually introduced and, naturally, one of these followed railway tradition, with a locomotive pulling a passenger coach. This coach or 'car' was often a 'double-decker', but sometimes several single-deck cars were hauled by one locomotive, in which case it was very difficult to distinguish between a tramway and a light railway. The second type of steam tram did not have a separate locomotive because the boiler and steam engine were built into the passenger car. Finally there was the tram with no visible power-unit at all – the cable car. Each car was connected to a cable, or wire rope, and hauled by a stationary steam engine.

Because of the difficulty of distinguishing between tram and train, it is almost impossible to say who invented the steam tram. During 1859 and 1860 several genuine trams, with their engines inside the car, were built in America. The first steam tram to operate in Britain was also of this self-propelled type. It was designed by John Grantham in 1872 and its two-cylinder engine plus twin Field boilers were built by Merryweather and Sons. The following year this tram made several trial runs in London, but it was not a complete success and had to be modified considerably, including a new single boiler by Shand, Mason and Co.

The first steam trams to operate a regular service in Europe were of the separate locomotive and car layout and began operations in Paris in March 1876. Their locomotives were built by Merryweather and Sons, who constructed over 170 tramway locomotives between 1875 and 1892. Locomotives were also manufactured by Henry Hughes and Co. of Loughborough, Kitson and Co. of Leeds and several other companies. The usual layout included a horizontal, railway locomotive-type boiler and two cylinders, about 9 inches in diameter and 15 inches stroke.

The tramway locomotive could be used as a direct replacement for the horse-tram; the same passenger cars could even be used. The cable tramway, however, involved a maze of underground cables and pulleys plus stationary engines and special cars which could be connected to the endless cable. Because of these complications, cable trams were usually only used in towns with steep hills. Edinburgh, for example, had cable trams in service from 1899 until 1923.

Most of the steam trams in Britain were replaced by more versatile electric trams during the 1900s, and it was during this same period that electrified railways really began to challenge the dominance of steam. Between 1903 and 1905 the District Railway of London's Underground was converted to electric traction, and in 1909 the electrification of suburban lines in South London was commenced. Even more advanced electric railways were being built on the Continent, where hydro-electric power stations were producing cheap electricity. In Switzerland, main-line electric locomotives were introduced as early as 1906.

An interesting locomotive was built by the Great Eastern Railway in 1902 to prove that a steam locomotive could match the performance of the new electric trains on suburban lines. In order to achieve a rapid acceleration it had three cylinders connected to ten driving wheels, and consequently became known as the *Decapod*. It created something of a record by accelerating a train from rest to 30 miles per hour in a mere 30 seconds. Unfortunately, it was said to be too heavy for certain bridges and the idea was dropped. One reason for the *Decapod*'s extreme weight was the fact that it was a tank engine and therefore carried some 1,400 gallons of water – plus all its coal – on the locomotive instead of in a tender.

By about 1900 the design of locomotives was so advanced that changes in the next, and final, fifty years were relatively minor when compared with the preceding half-century. The large 'Atlantic' type of express locomotive had been introduced in the United States during 1894 and in Britain four years later. This engine had a wheel arrangement of 4-4-2, which consisted of a four-wheeled bogie at the front, then four coupled driving wheels and a pair of trailing wheels under the driver's cab. The early 'Atlantics' had two cylinders, but later versions were fitted with three, and in most cases these were not arranged as a compound engine. Other locomotives were built with four cylinders, and one Midland Railway engine, built in 1908, even had eight.

Speeds at this time were very little slower than fifty years later, and in 1904 the Great Western Railway's *City of Truro* reached a speed of over 100 miles per hour hauling a light train. The famous 'City' class used steam at just under 200 pounds per square inch, but this was not superheated. In 1906 the G.W.R. became one of the first British railways to fit a steam superheater to the boiler of a passenger locomotive. The type of superheater installed was developed in Germany by Dr. Wilhelm Schmidt and patented in 1900. Superheaters were so successful that practically every designer in Britain adopted them within a few years. Feedwater heaters for the water supply to the boiler were also introduced by many designers during the early part of the twentieth century.

Locomotives incorporating Professor Stumpf's version of the Uniflow engine were built in Europe soon after 1908. The first British locomotive to be fitted with Uniflow cylinders was an express goods engine built for the North Eastern Railway in 1913. Several attempts to introduce Uniflow locomotives were made through the years, including one in 1927 which utilized steam at a pressure of

850 pounds per square inch instead of the usual figure of 200. But, in Britain, the Uniflow never seriously challenged the more conventional cylinder arrangements.

During the 1920s several important events in railway history took place. In Britain, probably the most noticeable change was the merging of the numerous railway companies into four large ones, which became the L.N.E.R., L.M.S., G.W.R., and S.R. This grouping took place from 1 January 1923 and lasted until the railways were nationalized in 1948.

A number of very large and powerful express locomotives were built during the period 1920 to 1930. Many of these had three or four cylinders supplied with superheated steam through piston valves operated by Walschaërts valve gear. To absorb the great power produced, six coupled driving wheels were fitted instead of four, and the 'Atlantic' wheel layout of 4-4-2 was superseded by a 4-6-2 or 'Pacific' layout. All the major British railway companies, except the G.W.R., adopted the 'Pacific'-type locomotive, which was rather strange because one of the first examples of this type was designed in 1908 by G. J. Churchward for the G.W.R.

Probably the most famous of the 'Pacific'-type locomotives were those designed by H. N. Gresley from 1922 onwards. These were very widely used by the L.N.E.R., and were later streamlined. It was a streamlined 'Pacific' called *Mallard* which, in July 1938, set up a new world record for a steam train, with a speed of 126 miles per hour. This record still stands and is not likely to be broken now that the steam era has ended.

The Great Western alternative to the 'Pacific' was a series of locomotives with a wheel arrangement of 4-6-0, and in 1925 one of their 'Castle' class locomotives was interchanged with an L.N.E.R. 'Pacific'-type. Each performed equally well, but the 'Castle' burned less coal. Of course, there were numerous other types of locomotives which are described in minute detail in many specialized railway books.

Efforts to increase the efficiency of steam-powered locomotives included the use of higher steam pressures, which led to the introduction of water-tube boilers, and steam turbines in place of the reciprocating engine. Turbines are the subject of the following chapter.

By 1940 diesel-engined locomotives were replacing steam engines on the railways of America and Europe. At sea there were very few ocean-going ships with coal-fired boilers and the steam turbine was replacing the reciprocating engine. The railways of Britain were in no hurry to change, and the nationalized British Railways began to introduce standardized steam locomotives for all their regions. But the higher efficiency of the diesel could not be ignored. From 1949 diesel locomotives began to appear on British lines and in 1960 the last British steam locomotive, the *Evening Star*, was built. This must have been one of the last examples of a large reciprocating engine powered by steam from a coal-fired boiler—the first was Newcomen's beam engine produced almost 250 years earlier.

Steam marine engines were built after 1960 as some overseas customers preferred this type of power, but the steam for these engines was usually supplied by oil-fired boilers. Although steam engine production virtually came to an end in the early 1960s the end of the era was still some years ahead, because many engines in service were still running satisfactorily. Even though an electric motor or a diesel engine is more efficient, and therefore cheaper to operate, a considerable sum of money is required for their purchase and installation. In the late 1960s steam engines were still working in factories, mills and workshops. They were also being used for pumping water at waterworks, winding men and coal up colliery shafts, and even for operating a swing bridge. But each year that passes sees the end of a few more examples and only a fraction of the total can be preserved.

The end of steam power on the railways of Britain—as far as most people were concerned—was the end of the steam age. A few stationary engines may still be working, but they are normally hidden from sight in engine-houses, whereas the well-loved steam locomotive could be clearly seen, heard and even smelt by all.

CHAPTER 13

Steam Turbines

Steam power for the twentieth century

Waterwheels and windmills were slow and cumbersome, but they were picturesque and they did provide power which could rotate a shaft. The majority of mechanical devices, from mill-stones to record-players, have to be driven round and round, and for this purpose the first steam engines had limited possibilities, because their pistons moved only to and fro. The difficulties of converting this reciprocating motion into a rotation, by means of a crank or Watt's 'sun and planet' gears, have already been described, as were the troubles encountered with reciprocating engines when run at high speeds. For many years, both before and after the invention of the steam engine, inventors strove to produce an engine which could compare with the simple waterwheel or windmill and provide rotary motion directly. These were known as 'rotary' engines, in contrast to reciprocating engines with a crank which were 'rotative'.

Hero's steam-driven 'aeolipile' and Branca's steam-wheel, described in Chapter 2, could certainly produce rotary motion directly, but the power they developed would have been negligible. However, they demonstrated the principle of the 'reaction' and 'impulse' turbine respectively, both of which were used for turbines driven by water during the early nineteenth century.

Encouraged by the success of water turbines, engineers went on to develop a steam turbine, but they encountered a fundamental difficulty. Turbines derive their power from a moving substance which has kinetic energy, determined by two factors—the weight of the substance together with its speed. Water, being heavy, can store considerable energy when travelling at moderate speeds, but steam is very light and must therefore travel very much faster than water in order to store the same energy. A turbine converts this store of energy into useful work, and so a steam turbine must rotate at speeds far in excess of its water-driven counterpart. In the mid-nineteenth century there were relatively few machines which required a very high-speed drive. Furthermore, gears, which could provide a reduction in speed, were still in the early stages of their development.

The turbine was not the only solution to the problem of producing a rotary engine, and an alternative was the 'steam wheel' of James Watt. In a letter of 1766 Watt refers to a 'circular steam engine' and for many years he divided his attention between this device and the reciprocating engine.

Basically the 'steam wheel', as Watt later called it, consisted of a tube bent into a circle and fitted with hollow spokes through which steam could be supplied to the annular space inside the tube. This annular space was divided into sections, by valves, and partially filled with mercury. The idea was to connect a steam supply to the section on one side of the mercury and a vacuum to the section on the other side. Steam tried to push the mercury through the tube, but, as mercury is a heavy liquid, it remained stationary while the wheel moved—or at least it was meant to. Although Watt built several experimental 'steam wheels' and details of one were included in his 1775 patent, these devices were never very successful. Once he had developed the 'sun and planet' gear and made his reciprocating engines drive a rotating shaft, the need for a rotary engine diminished.

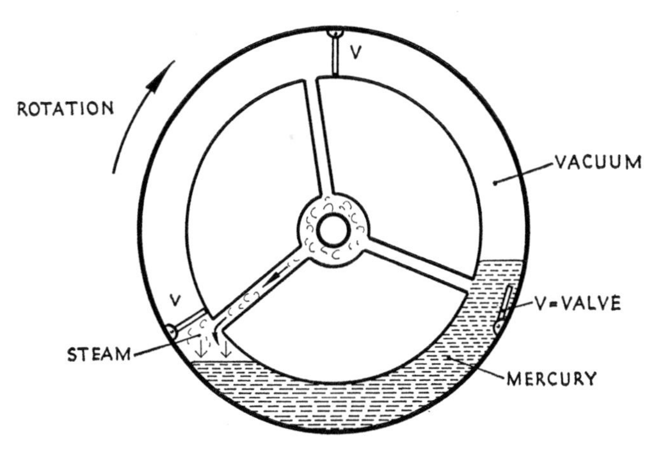

FIG 60 *James Watt's steam wheel of 1775.*

Several ingenious rotary engines were produced from Watt's time to the present century, but those using steam were generally small and of low power. Rotary petrol engines were very successful during the 1914–18 War as a source of power for aeroplanes. Although these were rotary engines, they were not true rotaries in

Watt's sense because they retained reciprocating pistons. Instead of a rotating crankshaft, they had a stationary one and the cylinders actually rotated around this. This type of rotary engine was even slower than a conventional engine and far more difficult to balance.

In 1784, only two years after Watt made his beam engines rotative, an experimental rotary engine was patented by Baron von Kempelen of Hungary. This worried Watt and he set to work investigating the new invention – in case it superseded his engines. At the end of his calculations he wrote:

> 'In short without god makes it possible for things to move 1000 feet pr." (i.e. 1,000 feet per second) it can not do much harm.'

Fortunately for Watt, he was correct. The new engine was a steam turbine of the reaction type, which followed in the pattern of Hero's 'aeolipile' and a water-powered turbine, invented by Dr. Robert Barker in 1743. All these turbines obtained their power from the reaction of a jet, but in the case of the steam jet it produced a high speed with very little power.

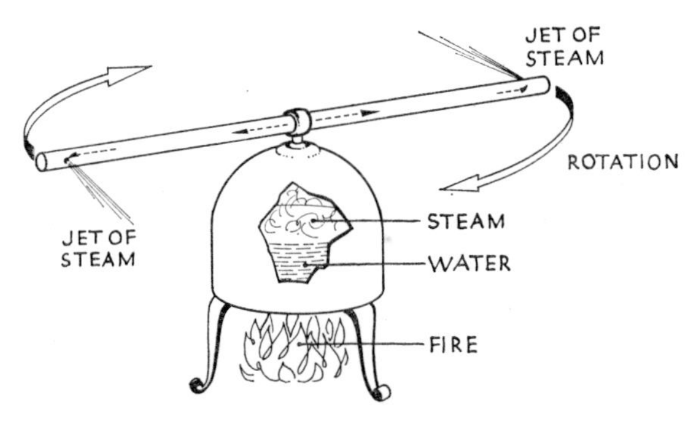

FIG 61 *Von Kempelen's reaction turbine of 1784.*

During the nineteenth century a number of attempts were made to design a practical steam turbine, and several famous engineers devoted some time to this problem. That prolific inventor Richard Trevithick patented a 'whirling engine' in 1815 and then proceeded to build one full-size. Steam jets were mounted at the extremities of arms, 15 feet in diameter, and the workmen thought it to be a device for hurling missiles at the French – so the story goes. Although it was, undoubtedly, impressive it was not a practical proposition, and Trevithick abandoned the project. Other engineers who have already been mentioned, including John Ericsson, James Nasmyth and Timothy Burstall, turned their attention to the problem of making a steam turbine. Neither these turbines, nor those built by several contemporary inventors, advanced beyond the experimental stage, and it was not until the 1880s that a practical working turbine emerged. When this happened two completely different designs were produced, one in England and the other in Sweden. The Hon. Charles Algernon Parsons (later Sir Charles Parsons) patented his first reaction turbine in 1884, while his Swedish counterpart Carl Gustaf Patrik de Laval was developing an impulse-type of steam turbine.

Charles Parsons was the son of the Earl of Rosse, a world-famous astronomer, who lived at Birr Castle in Ireland. Young Charles never went to school, but he received tuition from several famous men and was able to develop his natural skill with his hands in the workshops at Birr Castle. After studying at Cambridge University he served an apprenticeship with W. G. Armstrong & Co. at Elswick and spent two years working in an experimental workshop at Leeds. In January 1884 he joined Clarke, Chapman & Co. of Gateshead, near Newcastle, as a junior partner in charge of the newly organized electrical engineering department. The company produced machinery for ships, and to Parsons was given the task of developing a steam-driven dynamo to provide ships with electric light.

A dilemma faced Parsons. Existing reciprocating steam engines had sufficient power but were rather slow for a dynamo, whereas the experimental steam turbines were too fast and underpowered. The young engineer, taking a bold decision, set about making a steam turbine which would drive a dynamo directly, without any gears to reduce the speed. Before making his turbine, Parsons embarked on a series of experiments in order to solve a number of problems. One of these was the behaviour of shafts and their supporting bearings at very high speeds. Tests were carried out on shafts rotating as fast as 40,000 revolutions per minute.

Parsons decided to use the reaction type of turbine, but whereas Hero had fitted two jets from which steam escaped and 'pushed' the rotor round, Parsons devised a layout which had, in effect, over 1,000 jets. These were formed by blades around the periphery of twenty-eight rotating discs all mounted on the same shaft. High-pressure steam was allowed to escape through alternate rows of fixed and moving blades in such a way that its pressure dropped by a small amount at each 'stage'. When the steam entered the turbine, half went to the left, through fourteen stages, while the other half turned right, thereby eliminating any end thrust on the shaft. As the steam dropped in pressure through the multi-stages it increased in volume and, to compensate for this fact, the proportions of the steam passages increased towards the exhaust end of the turbine. Because the flow of steam was parallel to the axis of the rotating shaft, this layout was called an axial-flow turbine. In other words 'reaction' and 'impulse' describe the way in which the steam is used, but the terms 'axial' or 'radial' indicate the geometric arrangement for the flow of steam.

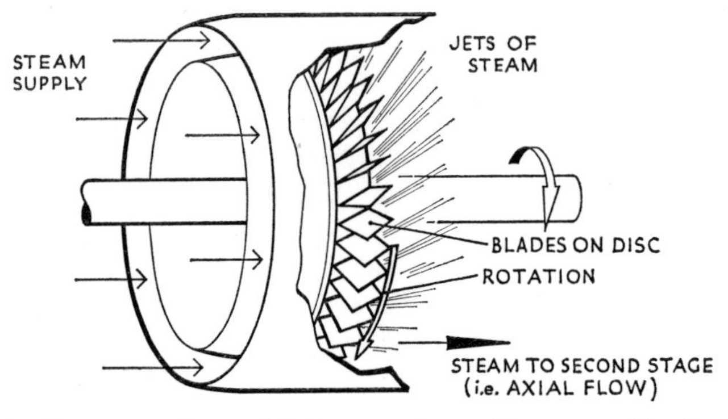

FIG 62 *One stage of an axial-flow reaction turbine, as used by Parsons. The steam supply is at a high pressure—not a high speed. It escapes as a high-speed jet which causes the disc to rotate.*

Despite its very short period of development, Parsons' turbine and dynamo were remarkably successful and the unit remained in service for about sixteen years. Steam at 80 pounds per square inch was supplied to the turbine, which drove the dynamo at a speed of 18,000 revolutions per minute and produced electrical power equivalent to about 5 horse-power. It was not as efficient as Parsons would have liked, and he immediately started work on improved turbines. His original example is preserved in the Science Museum, London.

De Laval started his experiments with a steam turbine at about the same time as Parsons, but his design took considerably longer to develop. Incidentally, this Swedish engineer had a French-sounding name because he was descended from one of Napoleon's soldiers who had fought for the King of Sweden and Norway. Like Parsons, de Laval started working on a reaction turbine, but eventually he changed to the impulse-type as illustrated by Branca. The Pelton wheel was a successful example of a water-driven impulse turbine—this had been introduced in the 1870s—and de Laval decided to make a steam turbine on the same principle.

Following the example of the Pelton wheel, de Laval used a single stage, consisting of curved blades (instead of buckets) radiating from a rotating disc, and the impact of a high-speed jet on these blades drove the turbine round. The difficulties of replacing water by steam, with the subsequent increase in speed, have already been mentioned, and these caused de Laval much trouble. To make his turbine economical de Laval had to use speeds around 30,000 revolutions per minute which led to difficulties with balancing, centrifugal forces, shaft vibrations and the lubrication of bearings. Yet even these were not the major problems. The speed was far too fast for ordinary machinery, and de Laval had, therefore, to design gears which would reduce it by at least ten times. There was also the problem of the jet of steam to be solved.

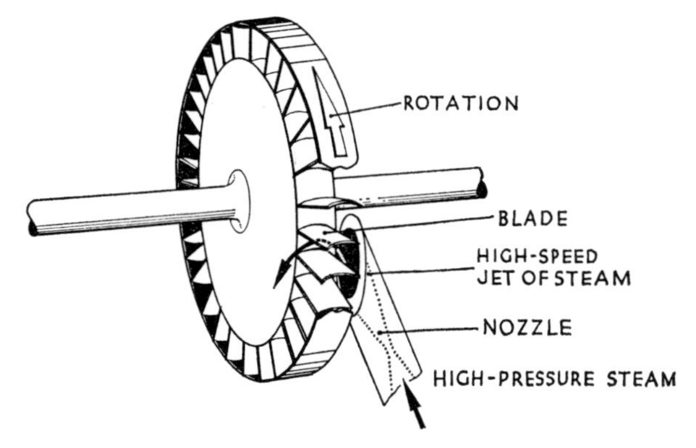

FIG 63 *The principle of de Laval's impulse turbine in which the impulse of a high-speed jet of steam on curved blades produces rotation.*

The jet of water for a Pelton wheel was placed in a tangential position which necessitated its being set some way back from the buckets: this was not efficient for the light-weight steam, and therefore de Laval positioned the jet on one side of the wheel at a slight angle. The steam impinged on the crescent-shaped blades and emerged on the opposite side of the wheel. By adopting this layout the angled mouth of the jet could be almost touching the moving blades, which travelled at a speed approaching that of a rifle bullet. At these speeds the design of a nozzle for the jet of steam was obviously important, and in 1889 de Laval made one of his greatest improvements with a new expanding nozzle. Water nozzles tapered down to their narrowest at the exit, but de Laval discovered that a higher speed could be obtained from steam by making the nozzle expand towards its exit. This difference was due to the fact that, whereas steam expands as the pressure drops through the nozzle, water does not.

During the 1890s de Laval increased the power of his turbines by adding several nozzles. Their popularity increased, particularly for smaller installations, and eventually they were built in Britain by Greenwood & Batley Ltd. of Leeds. Various sizes were available, ranging from a 1½ horse-power turbine with a 3-inch diameter wheel rotating at 40,000 revolutions per minute, to a 300 horse-power unit with a 30-inch wheel and a speed of 10,000 revolutions per minute. These speeds applied to the turbine itself, and usually reduction gears would be fitted to bring these figures down to a practical value.

While de Laval was developing his turbine, Parsons was improving the reaction turbine. In 1887 he designed a compound turbine with distinct high- and low-pressure turbines—each with many stages—mounted on the same shaft. As the steam expanded through the stages of the high-pressure turbine, the passage progressively increased, by lengthening the blades, until a practical limit was reached. Then came the low-pressure turbine with a considerably larger diameter and smaller blades, a combination which retained the same area for the annular steam passage at the changeover. Once in the low-pressure turbine, the steam continued to expand through the various stages until it was exhausted into the atmosphere. The names used for the various parts of a compound turbine are very confusing because compounding could infer just two stages, or two complete multi-stage units. These, in turn, might be called 'turbines' or 'cylinders'. In an effort to clarify the position and avoid confusion with piston-engine cylinders, it is assumed that a 'stage' consists of one moving row of blades (plus a row of fixed blades where applicable). Several stages make up a 'turbine' and more than one turbine constitutes a 'set'—as would a high-pressure turbine followed by a low-pressure turbine.

In 1888 Parsons designed the first electrical generating set to be driven by a steam turbine in a public power station. This unit was one of four installed in a Newcastle-upon-Tyne power station, but shortly after this success Parsons had a difference of opinion with his partners and left the firm. In 1889 he founded his own company

PLATE 39 *The steam turbine-powered* Mauretania *and the Forth Bridge. The dense black smoke suggests that her boilers were fired by coal—they were converted to burn oil in 1919.*

PLATE 40 (*Top*) *A large set of turbines being built by English Electric, Rugby.*

PLATE 41 *The shaft and blades of a low-pressure turbine—one of three for a set of Parsons turbines for the Pickering Nuclear Power Station, Canada.*

PLATE 42 *The port set of turbines for* Queen Elizabeth 2 *under construction. The high-pressure turbine is on the right-hand side (upper) and the larger low-pressure turbine is on the left.*

PLATE 43 *A turbine-powered locomotive of the LMS in 1935.*

PLATE 44 *Chapelcross Nuclear Power Station, Scotland, is similar to the first nuclear power station at Calder Hall. The four reactor buildings, each with two chimneys, and the four cooling towers are clearly visible.*

in Newcastle, C. A. Parsons & Co.; unfortunately, his patents remained the property of Clarke, Chapman & Co. Since these patents covered axial-flow turbines Parsons had therefore to start again with the alternative, radial flow.

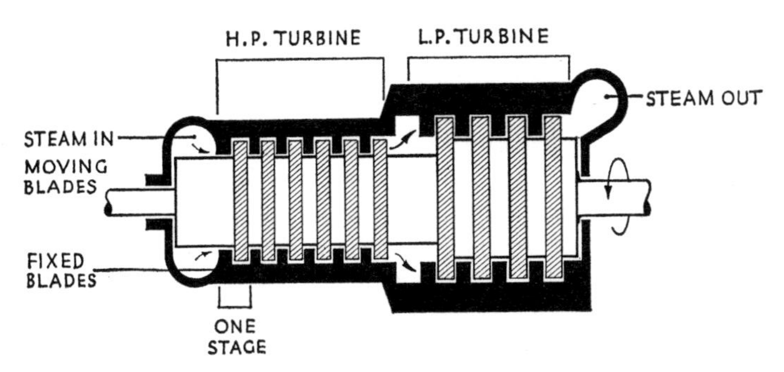

FIG 64 *A compound turbine set with a six-stage high-pressure (H.P.) turbine and a four-stage low-pressure (L.P.) turbine.*

Radial-flow water turbines (using the hair-drier principle) had proved very successful but, once again, the conversion to steam was not easy because so many more blades or vanes were required. A number of rotating discs carried the blades, which were arranged in concentric circles 'sticking out' from the face of the disc. Fixed blades were positioned between these rows of moving blades and the steam started from the centre, expanding through the various stages to the outside. Parsons overcame the difficulties and by 1891 he not only produced a successful radial-flow turbine but one with a condenser. Instead of allowing the steam to emerge from the final

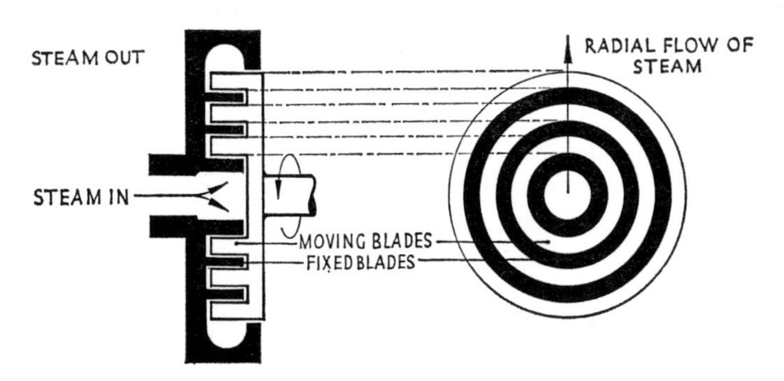

FIG 65 *The radial-flow principle.*

stage at atmospheric pressure, Parsons condensed the steam, thereby lowering the pressure well below that of the atmosphere. This in turn increased the drop in pressure through the turbine and gave an increase in power for the same amount of steam.

This condensing turbine, built for the Cambridge Electric Lighting Co., represented a very important step in the development of the steam turbine, because its steam consumption was less than that of a reciprocating steam engine of the same power. This was a remarkable achievement in view of the fact that it was only seven years since Parsons had made his first turbine, whereas the steam engine had been improved steadily over a period of nearly 200 years. The future of the new prime mover was assured because it was not only efficient, but also reliable, compact and smooth-running. Parsons' first condensing turbine is also preserved in the Science Museum.

Parsons was not satisfied with the radial-flow layout, and in December 1893 he persuaded his former partners to sell the original patent. Reverting to the axial-flow of steam enabled Parsons to design larger and more efficient turbines, although mechanical and constructional problems had to be overcome, particularly with the turbine blades. Orders flowed in from electric companies who were equipping their power stations with this revolutionary source of power. Some companies even replaced recently installed reciprocating engines with turbines, while other orders came from Europe.

Despite his success Parsons decided to expand his business, and in 1894 he turned his attention to marine steam turbines, setting out to build the fastest vessel afloat—no mean target for a new power unit. After building several models to determine the best shape for a hull, a full-size boat was built at Wallsend-on-Tyne. She was 100 feet long and powered by a radial-flow turbine driving a propeller shaft directly. The sleek *Turbinia* started her trials, but her speed was not as high as expected, and various types of propeller were tried, including several propellers on the same shaft. A test on the turbine revealed that it was delivering the correct power—the fault lay in the propellers. They were inefficient at the high speeds developed by the turbine because of a peculiar turbulence in the water, called cavitation. There was no easy solution, and in 1896 the original unit was removed and replaced by a set of axial-flow turbines which had three separate turbines each driving a propeller shaft. The high-pressure turbine was on the starboard side, with the intermediate one on the port side, and the low-pressure turbine, which was fitted with a condenser, in the centre. Also on the centre shaft was a fourth turbine to operate in the reverse direction when required, because turbines cannot be reversed. The turbines from *Turbinia* are preserved in Newcastle, at the Musuem of Science and Engineering.

For her trials in 1897, *Turbinia* was fitted with nine propellers—three on each shaft—and she attained the remarkable speed of 34½ knots, which was easily a new record. The *Turbinia* made a spectacular, but unofficial, public appearance at the Spithead Naval Review held in June 1897 to celebrate Queen Victoria's Diamond Jubilee. The little turbine-powered boat steamed between the stately lines of warships at top speed and thrilled the spectators. She must have impressed the Admiralty also, for Parsons received an order for a turbine-powered destroyer, H.M.S. *Viper*. The cautious authorities insisted that Parsons' company should deposit a sum of £100,000, just in case the *Viper* did not meet the specified performance. She achieved 37 knots during early trials, but unfortunately both *Viper* and her sister ship *Cobra* were later lost at sea.

While Parsons was developing the *Turbinia* he also fitted a turbine into a 22-foot launch, the *Charmian*. This small turbine, rotating at 19,600 revolutions per minute, was far too fast for a propeller and Parsons had gears fitted, to reduce the speed of the twin propellers to 1,400 revolutions per minute. A speed reduction of 14 to 1 was achieved by using twin gears with angled teeth in a 'herring-bone' pattern or, more technically, by a 'double helical reduction gear'. It had the advantage over ordinary straight gears of smoother running, because there was no sudden transfer of load from one tooth to the next. Later double helical gears were widely used for reducing the driving speed of steam turbines.

Parsons and de Laval were not the only engineers to design turbines, for experiments were being carried out in France, America and Switzerland. One of the major disadvantages of de Laval's impulse turbine was its very high speed of rotation. This resulted from a layout which converted all the pressure energy in the steam into kinetic (or moving) energy, which had to be absorbed by a single row of blades. August Camille Edmond Rateau of Paris decided to convert the impulse turbine into a multi-stage type, thereby reducing the speed. In effect, he passed the steam through a series of de Laval turbines, each with a set of nozzles and moving blades. Because the drop in pressure through each set of nozzles was only a fraction of the total, the speed of the jets—and hence the blades—was correspondingly lower. Rateau patented his idea in 1896 and built his first turbine two years later. In 1900 it caused considerable interest at an exhibition in Paris and, by 1908, Rateau turbines were being built under licence in Britain by Fraser & Chalmers of Erith and by Metropolitan-Vickers Electrical Co. of Manchester.

Rateau's system of separate nozzles for each stage was, in one respect, similar to Parsons' turbine and the multi-cylinder compound steam engine; in all three, the pressure dropped gradually stage by stage. The name 'pressure-compounding' was used to describe this arrangement, to distinguish it from another type of compounding possible with a steam turbine. The alternative was called 'velocity-compounding' and was derived from de Laval's impulse turbine, but instead of having one rotating disc of blades, the steam from the nozzles passed through alternate rows of moving and fixed blades. In this case the speed (or velocity) of the steam dropped gradually stage by stage, all the drop in pressure having taken place in the nozzles before the first stage.

The first turbine to include velocity-compounding was patented in 1896 by Charles Gordon Curtis, an American lawyer who became an engineer. In the following year, the General Electric Company took over the development of the Curtis turbine and after several years of hard work success was achieved. By 1903, the Curtis turbine was being built by several firms, including the British Thomson-Houston Co. of Rugby—a company associated with the General Electric Co.

During the early years of the twentieth century many different turbines were built, incorporating improvements in detail design and combining different types of turbine into one set. For example, Dr. Heinrich Zoëlly of the Swiss firm Escher, Wyss & Co. developed an impulse turbine in 1903 which was similar to Rateau's, but with blades of a different shape. Later Zoëlly designed a set of turbines for marine use with a high-pressure impulse turbine and a low-pressure reaction, or Parsons, turbine. Some very complex designs were produced by mixing the impulse and reaction principles together with the alternative methods of compounding and the two types of flow, axial and radial.

The success of axial-flow turbines, derived from the designs of Parsons and de Laval, rather eclipsed the radial-flow turbine in the early 1900s, but a Swedish engineer, Birgir Ljungström, aided by his brother Frederik, was studying this layout. When Parsons was forced to adopt radial-flow he mounted the concentric rows of moving blades on a rotating disc, while the fixed blades were fitted to the casing. Ljungström decided to make the 'fixed' blades move by mounting them on a rotating disc—in which case the two discs rotated in opposite directions.

By 1906 Ljungström was devoting a considerable amount of time to his contra-rotating radial-flow turbine and the following year he patented it in Britain. Technical and financial difficulties had to be overcome, but in 1910 the first full-size turbine was ready for testing. It developed 500 horse-power and ran under full load for several hours, rotating at 3,000 revolutions per minute. One advantage of this contra-rotating design was that it effectively halved the speed of rotation. Following the successful trials, orders followed, and in 1913 the Brush Electrical Engineering Co. of Loughborough started manufacturing Ljungström turbines.

Even while Ljungström was experimenting with his unusual turbine, the established designs were being used for some new and ambitious projects. A passenger vessel had been fitted with turbines and used on the River Clyde in 1901, but by 1907 two giant ocean liners were fitted with Parsons turbines. The sister ships, *Mauretania* and *Lusitania*, were built for the Cunard Steamship Company and represented the very latest in ship design, with a length of 780 feet and a tonnage of 38,000. Each was powered by turbines developing 70,000 horse-power and driving four propellers. A high-pressure turbine was directly coupled to each of the outer propellers, while the inner propellers were driven by low-pressure turbines. On her trials *Mauretania* covered 300 miles at an incredible speed of 27·4 knots (over 31 miles per hour) and soon became the fastest ship on the North Atlantic crossing—an honour she held for twenty-two years. *Lusitania* was less fortunate, being sunk by a German submarine during the First World War.

Mauretania's turbines required large quantities of steam at 195 pounds per square inch, and to cater for this demand she carried 25 Scotch boilers. Originally these were fired with coal, which accounted for the vessel's very large funnels, but in 1919 the boilers were converted to burn oil. After a long and successful life the *Mauretania* was broken up in 1935.

At the same time as *Mauretania* was being built, a super-battleship was under construction. When H.M.S. *Dreadnought* was launched it was said that her guns, armour-plating and speed made all other battleships obsolete. Her speed of 21 knots was achieved by fitting Parsons steam turbines—the first turbines used in a large warship.

Some ship-owners were wary of the steam turbine in these early years, and when Cunard's rival, the White Star Line, ordered two liners they compromised and used both reciprocating engines and a turbine. The *Olympic*, launched in 1910, and her sister ship the *Titanic* had three propellers: the outer pair were each driven by a separate triple-expansion engine and the one in the centre by a turbine. Steam from the Scotch-type boilers entered each high-pressure cylinder at 215 pounds per square inch and then passed through the intermediate cylinder, twin low-pressure cylinders, the turbine and finally the condenser. By the time the steam entered the turbine its pressure was only 9 pounds per square inch (which was below the normal atmospheric pressure of 14.7) and it left the turbine at only 1 pound per square inch. These liners were considerably slower than the turbine-driven Cunarders, their speed being 21 knots. In 1912, while on her first Atlantic crossing, the *Titanic* collided with an iceberg and sank, with the loss of many lives.

The system of a direct drive between a high-speed turbine and a propeller was suitable for some fast vessels, such as luxury liners and warships, but not for slower cargo ships. A reduction gear had been tried out by Parsons on his launch *Charmian*, and de Laval had made geared turbines, but the idea was not immediately developed because of the difficulties of manufacturing large, accurately shaped gears. Another method of obtaining a reduction in speed was patented in 1906, by Dr. H. Föttinger of Hamburg who devised a hydraulic drive. In its effect this was similar to a small high-speed water pump—driven by the steam turbine—pumping water into a larger water turbine which rotated at a slower speed and drove the propeller. This clever device was not an immediate success but was later used for some German naval vessels and the liner *Empress of Australia*.

Parsons carried out extensive tests to develop a system of gearing which would drive from the turbine to the propeller shaft and give a reduction in speed. In 1909 his Company bought an old cargo vessel, the *Vespasian*, in which they intended to install geared turbines and carry out full-scale tests at sea. Before making the conversion, the original triple-expansion steam engine was overhauled and tested, in order that any improvements could be measured when the turbines were fitted. Once these preliminary tests were completed, the engine was replaced by high and low-pressure turbines driving through helical gears on to the single propeller shaft, but nothing more was changed—the boilers, propeller and shafting remained the same. The gearing, which reduced the turbine speed to one twentieth, was a great success and the turbines used 15 per cent less steam than the original engines. The geared turbine was widely used from 1910 onwards.

Turbines for driving electrical generators were being built in large numbers, but the slower geared turbine opened up new possibilities on land as well as at sea. In 1910, C. A. Parsons & Co. built a turbine to drive a steel rolling mill for the Calderbank Iron & Steel Works in Lanarkshire. The 750 horse-power turbine rotated at 2,000 revolutions per minute while the mill was to rotate at only 70 revolutions per minute—a reduction of just under 30 to 1. This was too great for single-reduction gears and the reduction had to be made in two steps. The first helical gears reduced the speed from 2,000 to 375 and the second from 375 to 70 revolutions per minute. A giant flywheel was fitted, 23 feet in diameter and weighing 90 tons, to store up kinetic energy for the rolling process. Incidentally, when not highly loaded this turbine ran on steam from the exhaust of other engines.

Some very large turbines were built for power stations, and one of the largest in the years before the First World War was made by C. A. Parsons & Co. for an American power station. This huge axial-flow turbine was over 76 feet long with a height of 30 feet, and its largest blades were 19 inches long. It passed its acceptance trials in 1914, and when tested four years later it was found to be slightly more efficient—whereas most other engines deteriorate over the years, due, amongst other things, to wear in their cylinders.

During the period between the two World Wars the story of steam turbine development was one of steady improvement in efficiency and increased power. An average figure for the overall efficiency of a turbine installation in 1918 was 20 per cent, but by 1939 efficiencies had almost been doubled, and they were still rising. Because the efficiency depended on both boiler and turbine, high figures could only be achieved with the complex high-pressure boilers mentioned in Chapter 12. Although the steam turbine itself was relatively small, these boilers were so large that only power stations and large ships could accommodate them. In the early 1920s power stations were using steam at 300 pounds per square inch. A Babcock & Wilcox boiler for a Belgian power station, built in 1924, provided steam at 800 pounds per square inch, and another notable achievement was the Clyde steamer *King George V* with a boiler pressure of 575. This vessel, which can be considered to be a pioneer of very high-pressure marine turbines, was built by W. Denny & Bros. at Dumbarton in 1926.

Railway engineers were attracted by the high efficiencies of the steam turbine, and in 1921 a Swedish locomotive was fitted with a Ljungström radial-flow turbine. The performance was promising and several experimental turbo-locomotives were built in Britain from 1924 onwards, including one based on Ljungström's design in 1926. Probably the most successful was an L.M.S. express of the 'Pacific'-type built at Crewe in 1935. A Ljungström turbine was fitted which drove the wheels through a triple-reduction gear. Although this engine continued in normal service for a number of years the design was not adopted for further production, for reasons which were both complicated and debatable. One fact which should be remembered, however, is that, for a high efficiency, steam turbines have to be run at high speeds and under steady load conditions. These conditions can be achieved in a power station or by a ship on a long voyage, but a railway locomotive has to stop and start, accelerate and decelerate, climb inclines and run downhill. Attempts to make turbine-power more flexible in operation included hydraulic drives and even electric drives. In the latter case the turbine drove a generator which made electricity to drive electric motors, coupled to the wheels. Although this sounds rather complicated, it is quite feasible, and in 1938 the General Electric Co. of America built a turbo-electric locomotive.

An electrical drive was also introduced for marine use, and perhaps the most famous example was the great French liner *Normandie*, launched in 1932. Steam at 400 pounds per square inch, from 29 oil-fired, water-tube boilers, was supplied to four sets of Zoëlly impulse turbines, each with a high- and low-pressure unit. The turbines were coupled to electrical generators, and an electric motor drove each of the four propellers. On her trials in 1935 the *Normandie* achieved a speed of just over 32 knots, and later competed with the *Queen Mary* for the fastest Atlantic crossing and the coveted 'Blue Riband'.

Construction of the *Queen Mary* started in 1930, when she was ordered by the Cunard Line, but financial difficulties led to a halt in the work. However, when Cunard and White Star merged, and the Government gave financial help, work was resumed. The liner was launched in 1934 and she sailed on her maiden voyage in May 1936. *Queen Mary*'s turbines were supplied by 24 oil-fired, water-tube boilers delivering steam at 400 pounds per square inch—an installation very similar to the *Normandie*'s, but the *Queen* was fitted with Parsons combined impulse-reaction turbines and a single-reduction gear. Each of the four propellers was powered by a turbine set which expanded the steam in four successive turbines from high, through two intermediates, to low pressure—the equivalent of a quadruple-expansion steam engine. Four separate turbine-driven generator units supplied electric power for the engine room pumps and equipment, while three more provided power for the lighting and all the other services required by a floating hotel.

Many of the large liners which followed the *Queen Mary* were fitted with geared turbines, including *Queen Elizabeth* of 1938 and the post-war *United States*, *France* and *Queen Elizabeth 2*. The turbines for the latest *Queen* were designed by Pametrada, a name made from the initials of the Parsons and Marine Engineering Turbine Research and Development Association. This organization, with one of the longest-ever names, was founded by a group of companies in 1944. The *Queen Elizabeth 2* has two propellers and each is driven by a set of two turbines; a high-pressure impulse type is followed by a low-pressure turbine which divides the steam for double flow and expands it using a combined impulse-reaction design.

Although steam turbines are still being installed in large and fast ships they cannot compete with the economy of the diesel engine for many other vessels. Until a few years ago large marine diesels delivered up to 12,000 horse-power, and above this figure steam turbines were used. Recently, diesels of over 20,000 horse-power have been installed in large tankers: by comparison the turbines of the *Queen Elizabeth 2* deliver 110,000 horse-power. The limitation in the size of the diesel is principally due to the fact that it is a reciprocating engine with heavy pistons and connecting rods. Its basic layout is similar to the old marine steam engine but, of course, the diesel is an internal-combustion design and has a very high overall efficiency.

Invented about the same time as the steam turbine, the diesel was being installed in ocean-going vessels by about 1910, and from this time it developed very rapidly, particularly on the Continent. By 1920 it was twice as efficient as a steam turbine for medium-sized installations; or, in other words, the diesel consumed only half the amount of fuel-oil. Of course there were problems. The diesel engine was more expensive to manufacture and it was less smooth-running because of the reciprocating parts. Reliability, starting and servicing also caused a certain amount of trouble, but the diesel has been improved over the years until today it offers an increasing challenge to the steam turbine.

Strange though it may appear, many diesel-engined ships still require steam boilers, for, although a huge diesel engine can turn a propeller very efficiently, it can do very little else. Steam is supplied to small reciprocating engines and turbines, which drive water pumps, air-compressors, electrical generators and hydraulic pumps. These, in turn, power the engine accessories, steering gear, capstans and winches, as well as all the ventilation, heating and lighting equipment required to make the vessel habitable. Steamers may be slipping into history, but steam power at sea is still very much a current source of power.

CHAPTER 14

Jets and Atomic Energy

The future

Several major scientific achievements emerged from the Second World War and have been adapted for peaceful purposes with great success. Atomic energy is probably the most important, followed by the gas turbine engine, radar and rockets. The latter pair have had virtually no effect on the story of the steam engine, but the other two are important, one being 'for' steam power and the second 'against'.

The gas turbine, including its most famous variant the jet engine, soon became an established source of power, challenging steam and other engines. The gas turbine completed the quartet of important heat engines—the reciprocating steam engine, the reciprocating internal-combustion engine, the steam turbine and the gas turbine—which could also be called an internal-combustion turbine. A brief description was given in Chapter 1.

Experimental gas turbines were made as early as 1905 by H. Holzwarth of Mannheim who continued to develop the idea for many years. In 1920 he built a gas turbine to drive an electrical generator, but Holzwarth gas turbines were never produced on a large scale because their efficiencies were very low—a figure of 10 per cent has been estimated.

During the Second World War the gas turbine was produced as a practical power unit. It provided a high horse-power for a low weight which made it very suitable for aircraft, but it was still not efficient. To be efficient the gas turbine must operate at very high temperatures, which could be twice as hot as a steam turbine. Early jet engines operated at 800°C, and at this temperature the strength of steel drops to one fifth of its normal value. New materials have been produced to withstand higher temperatures, but the gas turbine is still in the process of development. At the end of the war prophets forecast that the gas turbine would rapidly supersede the steam turbine in power stations and become the universal source of power for all forms of transport. The process is proving slower than predicted, but whether this revolution will take place remains to be seen.

Today, in addition to the jet engine for aircraft, there are gas turbine installations in power stations, ships, lorries, trains, racing cars and even fire engines. Some of these are working examples, while others are merely experimental, and it would take an inspired prophet to forecast the trend in the next decade or so.

Because of the high-temperature problem, the gas turbine cannot yet compete with the steam turbine for continuous running. Where the former are installed in power stations, they are usually employed as a stand-by power supply to drive a generator when the normal generators are unable to meet a peak demand. The gas turbine has the great advantage that it can be started almost instantaneously—because it has no boiler to heat. On the other hand, the steam turbine is expected to run continuously night and day, for ten years or more, with an annual overhaul as the only break. Perhaps there is a place for both types of turbine in power stations.

The ever-increasing demand for power, and electrical power in particular, is threatening the world with a fuel shortage. It has been estimated that by the middle of the next century the world's supply of coal and oil will have been consumed—unless more atomic, or nuclear, energy can be used. Perhaps the most convenient method of converting nuclear energy into useful power is to produce electricity, and therefore one of the first tasks for scientists after the Second World War was to design a nuclear power station. Of course, there were many experiments and tests to be carried out, but in 1956 the world's first full-scale nuclear power station to produce electricity on an industrial scale was opened. This was Calder Hall Power Station, built on the coast of Cumberland, and it used uranium as a fuel—one ton of uranium in a nuclear reactor being equivalent to 10,000 tons of coal.

Deep inside the thick, protective walls of the Calder Hall 'reactor' atomic energy was released by splitting the core, or 'nucleus' of uranium atoms. Great heat was developed and passed on to carbon dioxide gas, which was pumped through the reactor to keep it cool. This reactor was therefore called a 'gas-cooled reactor'. The hot gas then passed into a large heat-exchanger and transferred its heat to incoming water before returning to the reactor. The water was converted into steam at high pressure and led away to drive steam turbines which in their turn drove electrical generators.

Disposing of the exhaust steam from the turbines of power stations can present quite a problem, not of a complex scientific nature but rather one of sheer bulk. To cope with all the steam,

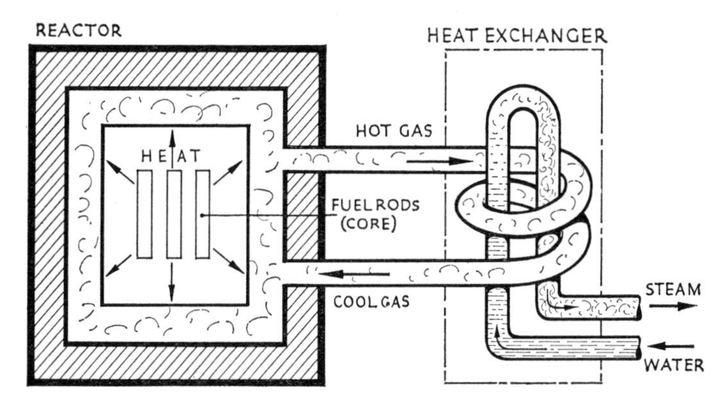

FIG 66 *A simplified diagram of a gas-cooled reactor.*

huge cooling towers have to be built. In these the steam rises, meets a spray of cooling water and condenses back into water—a far cry from the spray condenser used by James Watt. Because large quantities of water are required for the cooling towers, power stations are usually built on sites near the sea or a large river. Another waste product from a nuclear power station, which is even more difficult to dispose of, is the radioactive waste from the reactor. This deadly material will retain its radioactivity for many thousands of years and so has to be buried in expensive reinforced containers.

Even Calder Hall is now 'old-fashioned' and a new generation of reactors are in various stages of development. One of these is the advanced gas-cooled reactor which is similar to the one at Calder Hall, but the circulating carbon dioxide is heated to a higher temperature. The Americans favour a water-cooled reactor which in some cases produces steam inside the reactor, and this can be taken directly to the turbines—without the need for a heat exchanger. Liquid metal is also used to cool a reactor; for example the 'fast breeder reactor' at Dounreay in the North of Scotland is cooled by a mixture of sodium and potassium. It is 'fast' because the liquid metal absorbs heat from the reactor more quickly than a gas could, but a heat exchanger is required to produce steam. The fast breeder reactor has another interesting advantage; while in operation it makes (or breeds) fuel, but that is not all, for it actually breeds more than it consumes.

All the reactors mentioned so far obtain their energy by splitting an atomic nucleus—the principle of the atomic bomb. Scientists are now studying the even more powerful source of energy released by a hydrogen bomb, a thermo-nuclear process which also keeps the stars shining. But the day when a thermo-nuclear device will drive a steam turbine is a long way ahead. It may never come about because the turbine could well become obsolete. Methods of producing electricity without a turbine-driven generator are already in the experimental stage. The fuel cell is like a battery, and as long as it is supplied with oxygen and hydrogen it will not run down—but these gases are expensive to produce. Another device to produce electricity uses a process with the frightening name 'magnetohydrodynamics' or MHD. Very hot gases are moved at high speed down a tube between the poles of a powerful magnet. Wires are coupled to different parts of the tube and electricity is produced. This process, however, is still in its early stages because there are many practical difficulties. There are several other ways of converting heat into electricity, but as yet the combination of steam turbine and generator is the most efficient for large installations.

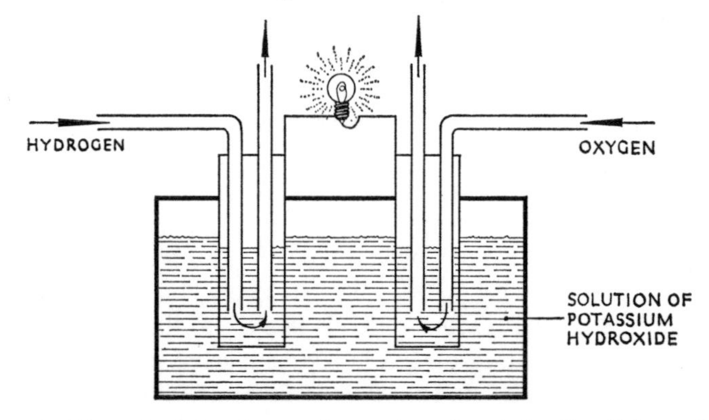

FIG 67 *The principle of an early version of a fuel cell which produced electricity.*

The small amount of fuel consumed by nuclear reactors made them very attractive to designers of vessels which remain at sea for long periods. Unfortunately, as so often happens, for every advantage there is a disadvantage, and in the case of a nuclear reactor one major problem is the size and weight of the protective material required around the reactor to prevent any harmful radiation from escaping. For this reason nuclear power is at present unsuitable for aircraft, locomotives and other land vehicles, but ships are able to carry the necessary screening.

The first working example of a passenger-cargo ship to use a nuclear reactor was the American *Savannah*, named after their pioneer of the steam age. Naturally this revolutionary ship was experimental and not expected to operate as a commercial success, but on the other hand she was more than just a floating laboratory. The *Savannah* could accommodate 60 passengers and her holds were capable of carrying almost 750,000 cubic feet of cargo. She could cruise at 21 knots and her fuel supply was sufficient for about three years.

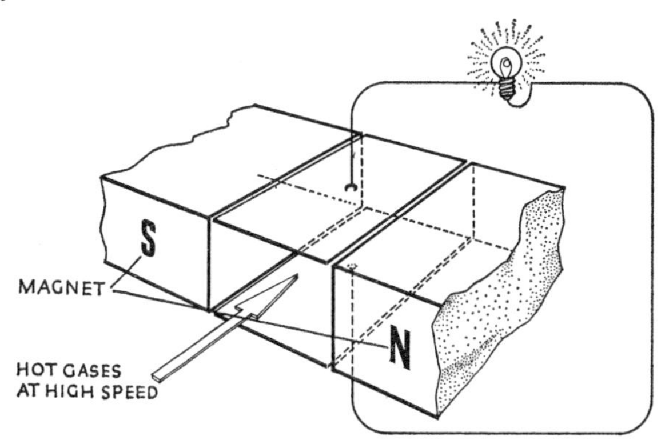

FIG 68 *Electricity being produced by an MHD generator.*

Savannah was equipped with a Babcock and Wilcox water-cooled reactor in which the cooling water did not turn to steam because the pressure was too high. The very high temperature water passed through a heat exchanger in which its heat was transferred to a second supply of water. This was circulated through a separate system of pipes and turned into steam at 465 pounds per square inch. Steam was supplied to a set of compound turbines, made by De Laval Turbine Inc. of New Jersey, which drove a single propeller through double reduction gearing.

The *Savannah* made many long voyages including several visits to Europe, and no doubt her owners acquired much useful knowledge.

A very different nuclear ship was produced in Russia at about the same time as *Savannah*. This was an ice-breaker called *Lenin* which was designed to cut through Arctic ice 6 feet thick. The *Lenin*'s ability to operate for over a year without refuelling was a great advantage when operating in the treacherous Arctic waters.

Like the *Savannah*, *Lenin* had a reactor cooled by high-pressure water, and steam turbines, but their transmission of power differed. *Lenin*'s turbines drove generators which provided electricity for electric motors attached to each of the three propellers. *Lenin* was a large vessel, with a displacement of about 16,000 tons and a speed of 18 knots.

Probably the most famous nuclear-powered vessels are the huge submarines which the navies of several countries have introduced. Using nuclear energy, these submarines can stay at sea almost indefinitely, remaining submerged for weeks at a time, but many details of their performance and power units are closely guarded secrets. Britain's nuclear submarines, which carry the Polaris ballistic missile, are 400 feet long and 30 feet wide. Their main power unit consists of a pressurized-water reactor and a steam turbine driving a single propeller.

Perhaps nuclear reactors providing steam for turbines will be the last example of steam power, but the end is not in the foreseeable future. It is over 250 years since the first working steam engine was built by Thomas Newcomen, with its tank of water over a fire in place of a nuclear reactor, and a slow, lurching piston instead of a smooth-running, high-speed turbine. Since that day in 1712, steam power has benefited mankind and it will continue to do so for many years.